California Natural History Guides: 10

EARLY USES OF CALIFORNIA PLANTS

BY

EDWARD K. BALLS

UNIVERSITY OF CALIFORNIA PRESS
BERKELEY, LOS ANGELES, LONDON

CALIFORNIA NATURAL HISTORY GUIDES

ARTHUR C. SMITH, GENERAL EDITOR

UNIVERSITY OF CALIFORNIA PRESS
BERKELEY AND LOS ANGELES, CALIFORNIA
UNIVERSITY OF CALIFORNIA PRESS, LTD.
LONDON, ENGLAND

LIBRARY OF CONGRESS CATALOG CARD NUMBER: 62-17531
PRINTED IN THE UNITED STATES OF AMERICA
ISBN: 0-520-00072-2 (PAPERBOUND)

7 8 9

CONTENTS

INTRODUCTION

In remote country regions of California it is still possible to find a few of the present-day native Indians using the old recipes and remedies derived from the plants they find growing around them. Modern packaged goods have reached the most distant corners of the state and have almost replaced the laborious methods of our predecessors. Within the past century this whole way of life has disappeared under the missionary zeal of modern civilization.

Information, however, on the uses to which the native California Indians and the early white settlers put the plants which they found growing around them is scattered through many books, journals, and pamphlets. Frequent inquiries for information on this subject from such groups as Scouts and schools have prompted the gathering together of the material here presented, in an attempt to describe some of the most important uses made in the early days, of plants of the state of California.

Before the coming of the Spaniards, the Indians gained their living from the countryside. The chief food along the coast, in the Sierra foothills, and in the central valleys was from the Oaks. On the deserts the main food sources were the Mesquite and Mescal (Century Plant). These foods were supplemented throughout the country by dry seeds from grasses and annuals, the few native wild fruits, various other nutlike seeds, and by the leafy parts of numerous plants. With the exception of the poisonous Zygadenes, all the bulbous plants were used for food. Along the coast immense amounts of shellfish were eaten, together with fish from the coastal waters, rivers, streams, and lakes. Game of all kinds was trapped or shot with bow and arrow. Birds were used for food only to a small extent. Rabbits were killed by the desert tribes with a club or throwing stick.

The two staple forms of food, atole and pinole, were used generally throughout most of California. Atole was the thick, souplike food made from ground, leached acorns or any of a number of other nutlike seeds. Pinole was the fine flour made by grinding seeds of Tansy-Mustard, Chia, or many of the grasses and flowering annuals. This was eaten either dry, pinch by pinch, or made into a mush with water. Bulbs, Mescal, and many other foods were cooked in the "earth oven," a pit dug in the ground, lined with rocks, and heated by building a fire in it. When the rock lining was thoroughly heated, the fire was raked out and the food was placed in the hot pit, covered with leaves, and then banked over with earth and sometimes more rock. For some foods a second fire was built on top of this oven but usually the food was left to cook in the heat stored in the rocks and earth for as much as twenty-four hours.

Cultivation of the soil was scarcely known. Some of the tribes of the lower Colorado River raised crops of corn, beans, squash, and watermelons. Toward the mouth of the river the seed of an unnamed wild grass was sown on the wet mud of the riverbed after the spring floods subsided. The seed was harvested when ripe, but no further attempt at cultivation was made. The introduced Barnyard Grass, *Echinochloa crusgalli*, was also used in this way to some extent by the Maricopa Indians. In the north the Karok Indians of the Klamath River region sowed Tobacco seed in prepared plots among the forests.

A large number of the native plants were used to cure the manifold ills, injuries, and ailments to which the people were liable. A wide variety of plants would often provide the remedy for a single ailment, meaning that if one cure was not available, others would probably be found at hand. The first Spanish settlers soon learned from the Indians the uses of some of the most important plants. These were the more readily adopted, in that the newcomers were prepared to make use of everything possible from their new surroundings. Naturally, in the course of a few years, as many of the introduced plants spread rapidly, they were absorbed into the Indian economy, providing further items in their list of possible foods and cures.

Before the arrival of the Spaniards, the clothing of the Indians was very simple, made from the skins of animals (chiefly deerskin): or from the fibers or leaves of such plants as the Yucca. Sandals, where worn, were made of twined fibers or more rarely from skins. Many, if not most, of the people went barefoot and almost without clothing.

A few of the eastern desert tribes learned the art of making pottery, and some stone vessels were used, but the majority of the population used baskets for all household, carrying, and storing purposes. The art of basket-making was very highly developed, and some of the finest known examples of baskets were made by the California Indians, particularly by the Chumash of the Santa Barbara coast and islands. Cooking pots and water containers were made of basketwork so finely and closely worked as to hold liquids satisfactorily. Because cooking baskets could not be placed on the fire, the technique of cooking was to heat rocks in the fire, lift them out with two sticks, drop them into the food in the basket, and then stir to prevent the heat from burning holes in the basket. This process was repeated until the food was satisfactorily cooked. Water was also boiled in this manner.

One early traveler reported that among the Coahuila Indians of the San Jacinto area a very roughly constructed loom was used for weaving rugs and horse blankets. It is affirmed that this was a native development and not borrowed from the Spaniards. But no other tribe in California is recorded as having advanced to this mechanical standard. Rabbitskin blankets were made with long strands of skin lightly twisted on Yucca fibers and sewed together. The woven baskets of the north were made on a foundation of withes with a regular twining action.

Few dyes seem to have been used, possibly because clothing was, in most tribes, at a minimum. Yellows, ochres, black, and red are the only colors referred to generally. Green was occasionally used in tattooing, and some blue from Delphinium was used by the Karok Indians in putting designs on their bows.

Indian homes were all of simple construction, mostly a framework of small poles or branches covered with a thatch of grasses, reeds, or leaves. In places this was again covered with earth or mud. Some of the huts were partly dug into the ground. In the forested areas in the north, boards were cut and used when fallen trees were available.

Over a large part of the country the sweathouse, or temescal, was one of the chief centers of village life. This was a place somewhat equivalent to the council house of many primitive peoples, where only the men gathered to smoke, discuss affairs, and sleep. The place was heated and was, in fact, a steam bath. It was not only a gathering place for the men, but a place where many ailments were treated with steam, medicinal herbs, and incantations.

The woman's place in the community was in the home and as the servant of the man. Hers was the job of providing the main foods. Fishing and hunting belonged to the men, but the women harvested the seeds, dug the roots, and gathered the herbs. Along the coast the gathering of shellfish was woman's work. The California Indians all lived in villages and were not nomadic or migratory as were many of the Indians of the Middle West. While some of them would move out to harvest Pinyon nuts, Mesquite, and Mescal, these moves were only for the short periods of the harvest.

Acknowledgments

For all the material here included my thanks are due to the authors whose works are listed as references at the end of this paper. My thanks are also due Philip A. Munz, Director Emeritus of the Rancho Santa Ana Botanic Garden of Claremont, California, for much help and good advice during the collection and presenting of the material, and for placing at my disposal the facilities of the library and for the use of many of the illustrations provided by the Botanic Garden; to Mrs. Gloria Day for considerable preliminary work on the subject which really formed the base from which the final work was developed; to Stephen Tillett and Richard Beasley for the excellent line drawings; and to Miss Gladys Boggess for much help as well as the arduous work of typing and checking manuscript. To Mr. Percy Everett, Superintendent of the Botanic Garden, I owe thanks for providing the impetus and encouragement which brought the work to completion.

A Note to the Reader

In the presentation of the material selected, it has been found desirable to include in a single statement all uses to which each plant was put. This poses the problem as to which category any plant may belong. Plants have therefore been placed under the heading in which their most important use falls, as in the case of the Yuccas where fibers were the primary product, but food, soap, and dyes were also obtained. In all cases plants are referred to by their common or vernacular names with the scientific name following the first mention of the plant. In the Checklist-Index at the end of the book will be found all the names of the plants mentioned in this work by common names in alphabetical order, each followed by the scientific name.

Oaks (*Quercus*)
With the exception of the deserts of southeastern California and the highest slopes of the mountains, Oaks are to be found throughout the state. Only the scrub Oaks were not generally used by the Indians, though even these were taken in times of acute shortage of other food. A. L. Krober states that "acorn soup, or mush, was the chief daily food of more than three quarters of native California."

Among the Live Oaks, the Coast Live Oak, *Quercus agrifolia*, was most highly prized. The other species of

Coast Live Oak

evergreen Oaks were gathered chiefly as an emergency food. Of the deciduous Oaks, the Valley Oak or Roble, *Quercus lobata,* was the first choice. It is said that the Indians of Mendocino County would gather as much as 500 pounds of acorns for a year's supply for one family. The Tanbark-Oak, *Lithocarpus densiflora,* was also very generally used.

The process of gathering, storing, preparing, and using acorns varied somewhat in detail from tribe to tribe, but the procedure was much the same throughout the state. The acorns were first slowly and thoroughly dried. They were frequently stored in their shells, though some groups shelled them before storing. Meal was always ground at the time it was to be used.

There are various accounts of the thorough leaching

Tanbark-Oak

process which had to be carried out in order to get rid of the bitter tannin properties. This leaching was generally accomplished after the meal had been ground, though records show that a few tribes sometimes buried acorns in swampy mud for a year. They came out purplish in color and ready to be roasted and eaten whole. This treatment was used among the Shasta Indians with acorns of the Canyon or Maul Oak, *Quercus chrysolepis*. Another treatment practiced by a few tribes was to leave the shelled acorns to mold in a basket and then bury them in clean sand in the river bed until they turned black. These were also ready for use without further treatment.

The process of grinding was simple though tedious. It was pounding rather than actually grinding, done in a stone mortar with a stone pestle which was often just a suitable rock used without any shaping, though there are many good specimens of tooled pestles which show that many of the tribes did fashion these implements. In the south the mortar was most frequently a worn hollow in a large rock near the gathering or drying grounds, though portable mortars were also used. Near to those acorn harvesting areas large rocks are found with numbers of these hollows, suggesting long use by many people. In the north the grinding stone was more generally a flat rock which had only a slight depression. Over this was placed a flaring basketwork hopper which was held in place by the stretched-out legs of the woman grinding (see Frontispiece).

Only a few acorns were ground at one time. The meal was winnowed or sifted over a circular piece of basketwork and only the finest meal was used. The coarser particles were swept up and reground until fine enough. Leaching was done in a "sand filter," either hollowed out in the sandy bed of a stream or in a large shallow basket, especially woven for this purpose. Water, hot or cold, was poured over and through the meal and the bitterness washed into the sand. Leaching with cold water might occupy the greater part of a day. The result was a substance very much like ordinary dough. The center of the dough, which would be free of sand, was reserved for baking into bread. The remainder which had some of the sand sticking to it was rubbed up with water in a basket and made into the soup called atole.

Watertight baskets were originally the only cooking vessels many of the Indians had, and the soup was heated by placing hot rocks in it (see p. 7). By some the heated rocks were dipped in water to wash off the ashes before they were placed in the soup. In the south a wooden "paddle" was developed for stirring.

The soup when cooked was usually a brownish red with a slightly sweetish but rather insipid taste. No seasoning was used. The soup was usually eaten by dipping one, two, or even three fingers into the basket which served as the common dish. Spoons of any kind seem to have been unknown to most of the Indians until after contact with the whites, though some of the coastal Indians of the north used mussel shells to scoop out their soup.

In baking bread the Indians of Round Valley in Mendocino County often mixed their acorn meal with a kind of red clay, stating that this sweetened it. The "pit oven" (see p. 6) was the usual Indian method for cooking acorn meal bread and a great variety of foods. When baked, the acorn bread was completely black and about like a soft cheese, but in a few days it became quite hard and then might have been mistaken for a piece of coal. It was quite sweet. Most of the Indians did not mix clay with their meal. It seems likely that most of the acorns eaten were in the form of soup rather than bread.

Although the most important use of the Oaks in California was for food, the bark had medicinal uses and was also used in curing hides and in making a dye. The bark of the Kellogg or Black Oak was much prized as fuel for parching seeds, as it burned long and slowly without a flame.

The galls which grow on the Valley Oak were collected for their juice, which was diluted and used as an eyewash. By putting rusty iron into it, a black dye was made which could be used in place of ink. The dye to blacken strands of Redbud, *Cercis occidentalis,* for patterns in baskets, was made by placing rusty iron in water in which the bark of *Quercus lobata* had been soaked.

California Buckeye (*Aesculus californica*)

The California Buckeye (see pl. 2) was used among many of the Indian tribes for food. It needed considerably more leaching than the acorns. The nuts were broken open and soaked in water for a day, then pounded into meal and leached in the sand filter. The process of washing and drying had to be repeated as much as ten times, count being kept by laying aside a stick for each time the washing and drying process was done. The "drying" between washings was rather a through draining, and the process could take a whole day. The mush was usually cooked as soon as the leaching was done, and was eaten right away.

The Indians of Round Valley cooked the whole nuts for as long as ten hours in a stone-lined pit, lined with willow leaves and covered with hot ashes and earth (see p. 6). After this the nuts were sliced or mashed and leached in running water for several days.

There are numerous references to the use of the Buckeye as a fish poison. Different tribes appear to have had varying practices in the use of this plant. The leaves and young shoots were used by the Concow and Yuki Indians. The more common reference is to the use of the nuts, which were crushed by stamping earth into them; the resulting mash was floated into small streams and stupefied the fish, making them rise to the surface to be taken in coarse nets or by hand.

Islay (*Prunus ilicifolia*)

The seeds from the Islay or Holly-Leaved Cherry (see pl. 1) were used, like the Buckeye, dried, ground into meal, and leached. This flour was not used for bread-making but only for the universal soup. The thin sweet pulp of the fruit does not seem to have been used in any way, though one reference states that the fruit, possibly the pulp, was used to make a beverage.

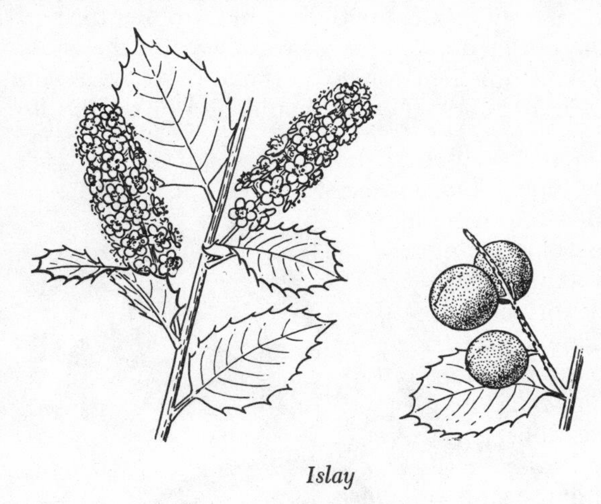

Islay

California-Laurel, -Bay, -Peppernut
(*Umbellularia californica*)

The thin-shelled nuts of the California-Laurel, -Bay, or -Peppernut were used in some quantity and often stored for winter use. When needed, the nuts were parched or roasted in the ashes of a fire, then cracked and eaten. There is record also of their having been ground and shaped into small cakes after roasting and used as "bread." The bitter quality was dispersed by the roasting.

The oily, pungent leaves were used in many ways medicinally. Headache was cured by placing a piece of the leaf inside the nostril or by binding a number of leaves on the forehead or under the hat. For chronic stomach ailments a large quantity of the leaves was bound round the body and left for several days. A tea made from the leaves was taken as a cure for stomach pains and headache. To cure rheumatism the Bay leaves were used in a hot bath or in the steam bath. The oil caused the skin to smart and made a thorough rubbing necessary. The white settlers adapted this by combining the oil from this leaf with lard and rubbing the body with the ointment. The leaves were used in the Indian houses as a repellent for fleas. When colds were common, boughs of the Laurel would be put on the fire,

California-Laurel, -Bay, -Peppernut

and the smudge produced was allowed to "fumigate" the house.

Mescal, Century Plant, Agave (*Agave*)

There are several different species of Mescal which grow over a large part of the desert regions of the southwestern United States, chiefly in the rocky and mountainous areas at altitudes ranging from 1500 to 5000 feet. *Agave Shawii* (see pl. 2) was to be found, rarely, along the coast of southern San Diego County. It is still abundant in Lower California. Within the boundaries of California only two species of Mescal are at all frequent in the areas where the Indians made use of

Mescal, Century Plant, Agave

them. Both *Agave deserti* and *Agave utahensis* var. *nevadensis* were referred to as Mescal.

The most important use for these plants was as food. In spring, when the flower buds were just beginning to shoot up from the center of the rosettes, the Indians came in large numbers to the places where the Mescal grows abundantly. Here they camped for weeks at a times, gathering, cooking, eating, and preserving the Mescal heads. The buds were cut out from the plant crowns with a long lever of stout wood, beveled at one end, which was tapped sharply with a stone to cut the crisp "cabbage" loose. These heads were collected in huge quantities and the cooking became a communal project.

The pit or oven in which the cooking was done (see p. 6) was about one-and-a-half to two feet deep. Great care was taken in selecting the fuel so that none was used which would give the finished product a bitter or undesirable taste. When the remains of the fire were raked out, a layer or covering of Agave leaves or grass was placed over the ashes and the Mescal heads were laid in the "oven" in layers, finally covered with another layer of leaves or grass, banked over with sand or earth, and left to bake until the following day. In some places these baking pits were as much as nine feet in diameter, and were used repeatedly season after season. A considerable rubbish heap often accumulated nearby from the cleaning out of the pits and the discarded fire-cracked stones which had to be replaced each time the "oven" was heated.

The baked heads, when taken from the pit, were stripped of the charred outer leaves, leaving a brown juicy mass, very sweet and nutritious. A large quantity of this was eaten just as it came from the baking, but much was also worked into cakes, dried, and thoroughly redried for winter storage. This dried product was as important an article for bartering with the neighboring tribes as was the very similar article made from Yucca fruits.

The preserved Mescal was eaten dry or recooked and used with other foods. A good, sweet drink was made by boiling pieces of the cake in water. In preparation and uses, Mescal and Spanish Bayonet were nearly identical. The California Indians do not seem to have made any use of alcoholic drinks prepared from either Mescal or Yucca, as did some of the tribes in Arizona, New Mexico, and Mexico.

The young flower-stems and unopened flower buds were cut into short lengths and roasted with the Mescal heads. The golden flowers were boiled, dried, and preserved for winter use. They would keep in this condition for some years and were still sweet and palatable when recooked.

The very young leaves were occasionally eaten raw, and the ripe seeds were ground into flour, often mixed with the meal of the Mesquite, and used as a gruel or pinole.

Mescal was of almost equal importance with the Yuccas in the production of fiber in some areas. The fiber was extracted from the dry leaves by beating, and from the fresh leaves by soaking and rotting off the pulp and outer skin, in much the same manner as the Yucca fibers were handled. Some tribes, after eating the roasted Mescal, saved the fiber and washed, cleaned, separated, and twisted it into cords. String of this fiber was usually rolled on the bare thigh by the women. A fairly common practice with the fibers from the Mescal heads was to clean them and tie them into bunches which served as hairbrush and comb. The dead leaves contained the stoutest fibers, and varying thicknesses of cord or rope were made by twisting together two or more strands of the rolled cord. Most of the uses to which Yucca fibers were put were also served by the fibers from the Agaves. Of cords "rolled on the thigh," David Prescott

Barrows says, "Such cordage is of the strongest. Bowstrings made from it last for years." From this cord the Indians also made carrying nets and baby hammocks.

The women of several of the southern tribes made use of charcoal from burned *Agave* for tattooing by pricking in the bluish-black patterns with a thorn of cactus, *Opuntia*.

Mesquite (*Prosopis juliflora* var. *Torreyana*)

The Mesquite is a common shrub, often a small tree, through the washes and low places of the Colorado and Mojave deserts. It spreads eastward as far as Texas and into Lower California. For the Indian tribes living on the desert it was perhaps the most important plant, providing food, housing, and to some extent, clothing.

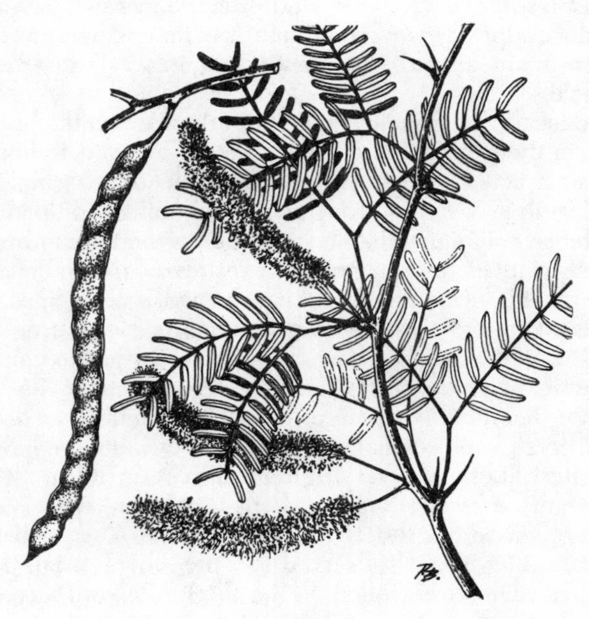

Mesquite

The long yellow beans were collected in very large quantities when ripe. They were stored in coarse, open-work basket "granaries," sometimes without drying, but more often dried thoroughly before storing. Generally the Mesquite beans were ground to a coarse meal and mixed with water without removing the seeds. This mush was eaten after being left to stand for a few hours, and the slight fermentation which set up in it was thought to improve the taste. Among some tribes the seeds were taken from the pulp and ground into flour which was made into pinole (see p. 6), and there are a few records of bread made from these ground seeds.

Great quantities of the beans were eaten fresh as they were gathered. The meal made from the pods was never cooked. Only in boiling the whole beans to make a drink, or in baking bread from the ground seeds, was the Mesquite cooked. The pulp of the beans is a very sweet and nutritious food, and the common drink made from freshly ground Mesquite was in general use wherever the plant grew. The beans were ground in stone or wooden mortars or holes in the ground. Sometimes a woven basket lined the hole. One tribe is said to have ground its Mesquite in hollows in the earth, mixing soil with the meal to sweeten the flavor. (For a similar practice with acorn meal, see p. 12.) Wooden mortars, used by some of the desert Indians, were usually made from a log of Mesquite wood with the hollow burned out. Long slender pestles of wood or stone were used to crush the meal. The seeds were broken or ground on a metate, usually a flat piece of rock, used much like a pastry board. The crushing was done by either a back-and-forth, rolling motion or a circular, grinding motion.

Mesquite wood was almost as important to the desert Indians as the fruit. Poles for building their homes and wood for making what furniture was used, mostly stools, came from the Mesquite. Charcoal was burned from this wood, and it was considered the best firewood because it burned long and slowly and gave great heat.

A Mesquite fire was usually used to bake pottery. The pots were lifted out of the fire with two sticks and filled, while still hot, with a fine gruel of ground Mesquite beans to make them non-porous. The charcoal from Mesquite was used in tattooing, giving a blue color. Fibers from the roots, bark, and inner bark were used by several tribes in making their beautiful baskets. War clubs and spears, bows, and the shafts of arrows were all made from this wood. The points for arrows were often made from Mesquite wood hardened in the fire. Throwing clubs, used especially in hunting jackrabbits, and digging sticks were also on the list. The digging sticks were about four feet long and two inches in diameter, sharpened to a point by burning and hardening in fire. These were used by the women in collecting bulbs for food.

A clear gum which exudes from cuts in the trunk of the tree was pleasant to eat. It was also used as a glue. Dissolved in water, it was used to ease sore throats and as a lotion for sore eyes. This gum was boiled to make a black dye for painting patterns on pottery. A blue dye was also obtained from the leaves and the fruit. The boiled gum, mixed with mud, was plastered on the hair by both men and women and left for a day or two. When the "pack" was washed off it left the hair black and glossy, and the lice, which were only too frequent, were all removed. This treatment was used particularly for gray hair which it dyed black. In some areas trees were owned outright by individuals and were so marked, usually with a bunch of Arrowweed. Wild plants on unclaimed ground were regarded as belonging to the people who were living nearby, and others were expected not to gather the crop from those trees.

Screwbean (*Prosopis pubescens*)

This shrub (see pl. 3) is not nearly so widespread as the Mesquite to which it is closely related. It was therefore not nearly so important to the Indians as the Mesquite, though it was used in very much the same ways. When first gathered, the fruit was sour or bitter, but after treatment the beans, which are smaller than those of the Mesquite, were much sweeter. In most tribes the Screwbean was "cooked" by placing the harvest in a pit which was lined with the green leaves of Arrowweed, *Pluchea sericea.* In some cases the pits were fifteen feet across and four or five feet deep. The beans were generally placed in layers between leaves of Arrowweed or Cocklebur, *Xanthium.* Sometimes water was sprinkled on the beans and when the pit was filled the whole was banked over with earth and left to cook. No fire was used. It is hard to account for the difference in the length of time which was given to this "cooking." The Pima Indians left the beans for "three or four days." The Mohave left them for "about a month." After treatment the beans were taken out and dried for storage. It is said that these beans were much more likely to become "wormy" in storage than the Mesquite, but the worms were ground up with the beans and used with the meal. A fermented drink made from the meal of the Screwbean was used mostly by the tribes living along the Colorado River.

By boiling down, a fair substitute for molasses was obtained from the Screwbean. Little use was made of the wood of this shrub. The Coahuila Indians made a hunting bow from it. The Mohave Indians stripped the bark which they used for binding pottery.

Chia (*Salvia Columbariae*)

Chia was one of the most widely known of the native plants used by the Indians of the West Coast. This annual grows over a very large part of California in dry open places below an altitude of about 4000 feet. Often it is found in very large numbers. Chia seems to have held an almost unique place in the economy of the Indian tribes of the south and it was not long before the Spanish settlers came to know its uses. The Mexicans also used it extensively. In 1894 the seed of Chia was sold in stores in Los Angeles at from six to eight dollars a pound. The small, slippery, gray-brown seeds were gathered in very great quantities by the Indians, who harvested the dry seeds from this and many other annuals and grasses by bending the heads of the plants over a rather flat, tightly woven basket and beating the seed into it with a basketwork fan or paddle. It is told that the men cut and bundled the plants of Chia when the seed was ripening and brought them into the village where they were laid on a patch of hard-tramped earth and the seed threshed out with rods. The women then winnowed out the seed. This was not a common prac-

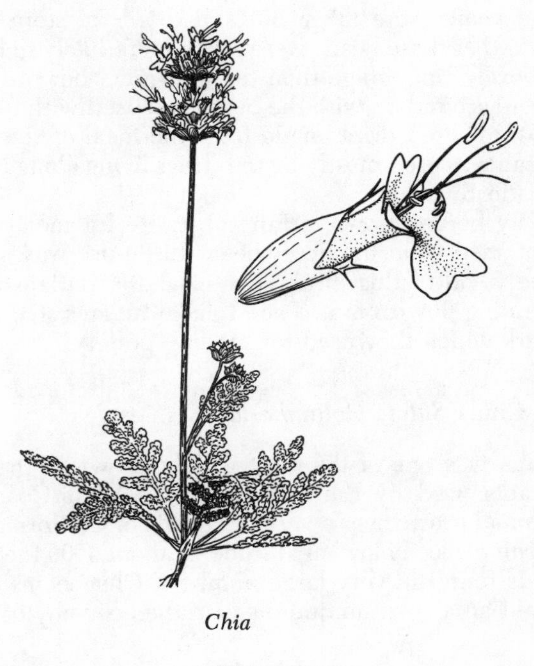

Chia

tice, for it was the woman's place in all the tribes to bring in the seed harvest.

Chia was used in many different ways. Seeds dropped into water and stirred made a slightly flavored soft drink. More generally, the Indians ground the dried or roasted seed to a fine flour. This was sometimes eaten dry in pinches, diluted to a drink, or more often mixed with water to make a kind of gruel. The meal contains a great deal of glue which swells in water. It has a very high food value and is easily digested. One teaspoonful of the seed was said to be sufficient to keep an Indian going for twenty-four hours on a forced march. For desert travel, a few teaspoonfuls in a quart of the warm alkaline water from desert water-holes is said to neutralize somewhat the dangerous qualities and make a refreshing and nourishing drink.

The Spanish settlers, when they used Chia to make a drink, often added a little sugar and some flavoring such as orange or lemon. Sometimes the Indians baked the flour into little cakes or biscuits. In any form it has a nice nutty flavor. The meal was often mixed with corn meal or wheat flour or with the meal of other seeds, largely to give it special flavor.

Because of its glutinous properties this seed was sometimes used in the place of linseed. A couple of seeds placed under the eyelids at night helped to clear the eye of any inflammation.

The seed of the very handsome Thistle Sage, *Salvia carduacea*, was also used for the same purposes and sometimes under the same name, Chia. These seeds are twice as large as those of *Salvia Columbariae* but are very much like it and have the same qualities.

Tansy-Mustard, Peppergrass
(*Descurainia pinnata*)

This weedy annual is very widespread throughout a great part of California. In some areas it grows in considerable numbers. The small red seed was gathered by many of the Indian tribes and used for food. Generally it was parched by tossing in a basket with hot stones or live coals. It was then ground to a fine flour on a metate or in a mortar and made into mush. The Indian name for the flour, pinole, comes from the Aztec original, pinolli, meaning "seed flour." Because of its peppery taste it was often mixed with the flour of other seeds to make a more palatable food. The young leaves were sometimes boiled or roasted between hot stones and eaten as green vegetables.

The ground flour was used for poultices, and a tea made from it was used for summer complaints, especially for children.

Where the Mescal was in common use, a drink was made by squeezing the juice from dried crowns of the Agave and mixing the juice with the flour of the Tansy-Mustard.

Tansy-Mustard, Peppergrass

The seeds of several other members of the Mustard family were used in these ways by the Indians. Even the introduced weed, Shepherd's Purse, *Capsella Bursa-pastoris,* was used to some extent in making pinole. On the deserts another Peppergrass, *Lepidium Fremontii,* was used, often with other seeds, in gruel and bread. *Caulanthus crassicaulis,* one of the many plants known as Wild Cabbage, was used more as a vegetable, though the seeds were also ground into meal and used as pinole. The young leaves were gathered and dropped into boiling water. After a few minutes they were taken out, washed in cold water, and squeezed. This was repeated five or six times to get rid of the bitter taste and harmful salts. The young leaves of the Prince's Plume, *Stanleya elata* and *Stanleya pinnata* (see pl. 1), were used in the same way. After the final washing the remains were often dried to be boiled again when wanted. A number of other plants belonging to this family were also used in the same ways.

In the coastal regions the seed of the attractive Lace-Pod, *Thysanocarpus curvipes* var. *elegans,* was used in

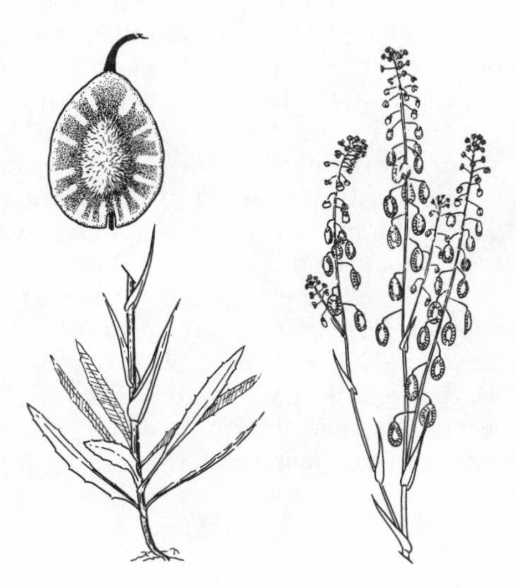

Lace-Pod

the same way as the Tansy-Mustard and Peppergrasses, for mixing with other seed flours in the making of gruel and soup. A tea made from the whole plant was sometimes used to cure stomach-ache, or a drink made from the leaf was used to relieve colic.

Pinyon (*Pinus monophylla*)

Several species of Pine are know by the name Pinyon, but in southern California the most common is *Pinus monophylla,* the One-leaved Pinyon (see pl. 3). It is to be found on dry rocky slopes and ridges from 3500 to 9000 feet along the east slopes of the Sierra Nevada, in the mountains bordering the Mojave Desert, and in a few places along the west base of the Sierra Nevada north to Tuolumne County.

This tree provided one of the most important items of food for many of the Indian tribes, and whole villages traveled many miles to harvest the seeds. Often the unopened cones were beaten from the trees by men with long poles. They were gathered up by the women and children, piled into heaps, and set on fire to burn off the pitch which is generally so abundant on the green cones. The seeds were then easily separated from the charred cones. The shells were cracked by rolling a stone pestle lightly over them on a metate. The husks and any other debris were cleaned out by winnowing. After the cones began to open on the trees the children would gather the nuts from the ground while the women took them from the trees. The seeds were parched to preserve them for a longer time. There is no reference in the literature to any form of preparation made from these kernels; they were eaten dried or roasted. They are still much sought after.

In the areas where they are abundant, the seeds of all the large-seeded Pines were used as an important item of food. The Coulter Pine *Pinus Coulteri,* was used in the south; the Digger Pine *Pinus Sabiniana,* in the foothills and the Coast Ranges on either side of the San Joaquin Valley; and the Sugar Pine, *Pinus Lambertiana,* throughout its wide range from southern California to Oregon.

Much less generally known are some of the other uses served by the Pinyon Pine. The gum or resin which drips from cuts, burns, or broken branches was collected in quantity and used medicinally. Chewing the gum was soothing to a sore throat. Sometimes the resin was dried, powdered, and applied to the throat with a swab. Boiled to make a hot tea, usually with parts of some other plant such as small twigs of the Utah Juniper, *Juniperus osteosperma,* it was taken as a cure for colds. This resin potion was also used in different tribes as a cure for rheumatism, tuberculosis, influenza, and chronic indigestion. The heated resin was used as a dressing to draw out embedded splinters or to bring boils to a head. It was sometimes used with crushed plants of Turtle-Back, *Psathyrotes ramosissima,* or the finely chopped tips of twigs of the Utah Juniper for the same purposes. The hot resin dressing was also used for sores, cuts, swellings, and insect bites. Smeared on a hot cloth the hot resin was used much as a mustard plaster in treating pneumonia, sciatic pains, and any general muscular soreness.

The pitch or resin, from most of the species of pines, served some tribes as a cement or glue, to some extent for chewing gum, and also in healing burns and sores.

The gum from the Sugar Pine was one of the most sought after. It is very sweet (from which comes the name). It is also a mild laxative and as such was well known, especially by the northern tribes. Dried and powdered, the gum was used to cure sores and ulcers.

The Karok Indians used the seeds of the Digger Pine to decorate dresses worn in their dances. For stringing the beads, holes were made by rubbing the ends of the nuts on rough rock. They were strung on thread made from Iris fiber.

Mariposa-Lily, Sego Lily (*Calochortus Nuttallii*)

This beautiful-flowered bulb is widespread from 5000 to 9000 feet on dry brushy slopes and flats east of the Sierra Nevada from Inyo County to Oregon and Montana. It is one of a group of western bulbs which are generally known as Mariposa-Lily or Butterfly-Tulip (see pl. 1). While the bulbs of all these Mariposas were used for food by the Indian tribes of the areas where they grow, the Sego-Lily was also adopted by the early Mormon settlers as well as by hunters and miners. The name Sego-Lily, which has also been debased to Sago-Lily, comes from an Indian name for the plant and applies particularly to *Calochortus Nuttallii.* The bulbs were gathered in enormous numbers. The only implement the Indians had for this job was

Mariposa-Lily, Sego-Lily

the digging stick, a stout stick sharpened to a point or bevel and hardened in the fire. The harvested bulbs were cooked in the rock-lined earth oven (see p. 6). Other Mariposas which grow on the western side of the Sierra Nevada, such as *Calochortus luteus, Calochortus venustus,* and *Calochortus macrocarpus,* were all used in this way.

Some of the smaller *Calochortus,* often known as Star-Lilies or Fairy Lanterns, were eaten raw because they were small and sweet.

Camas, Quamash (*Camassia quamash* ssp. *linearis*)

A bulb from wet meadows below about 2500 feet, the Camas is widely distributed through the Coast Ranges from Marin County northward to south Washington and in the Sierra Nevada from El Dorado County northward. This food item was undoubtedly the most important bulb to the native Indian population of the West Coast. It is often found in great quantities, and the bulb is much larger than most of the other native bulbs such as Wild Onion, *Allium, Calochortus,* or Indian Onion, *Brodiaea.*

The bulbs were usually cooked in the stone-lined earth pit where they baked for twenty-four hours or more. When taken out, they were soft and darkish brown in color and very sweet and nutritious. The cooked bulbs were either eaten straight from the "oven" or could be pressed into cakes and dried in the sun, and were thus preserved in the same manner that the Indians of the south preserved the Mescal. The cooked bulb is said to have the flavor of chestnut.

Tule-Potato (*Sagittaria latifolia*)

The common Arrowhead, Tule-Potato, or Wappato, is spread through a large part of California below 7000 feet, on the edges of ponds and in slow streams. It is an extremely variable plant. The roots (tubers) of this

Tule-Potato

attractive water plant were much used for food by both the Indians and the Chinese. The harvest was usually made in late summer as the stems and leaves were dying. Indian women, wading in the water and pushing small canoes before them, loosened the tubers with their toes so that they floated to the surface and were gathered into the floating baskets. They were baked in the embers of the fire, skinned, and eaten either whole or mashed. Lewis and Clark reported the use of this root for food on their famous expedition to the Northwest.

Cat-Tail (*Typha latifolia*)

In the days of the Indians much more use was made of the Cat-Tail, *Typha latifolia,* than in the present day. A kind of bread was made from the pollen, the thick, creeping root was roasted or dried raw and ground into meal and the young shoots were eaten as Bamboo shoots and many other young growths are used today. Floor mats and roofing thatch were made from the leaves, which with the leaf-sheaths, were used for calking materials in canoes and houses.

Jojoba, Goatnut (*Simmondsia chinensis*)

A common shrub on dry, rccky hillsides below 5000 feet on the western Colorado Desert, south of the San Jacinto Mountains to San Diego and to Arizona, Sonora, and Lower California, the Jojoba has large nuts rich in oil, which were eaten with avidity and without preparation by Indians, children, sheep, and goats. The Coahuila Indians, in whose territory they grew abundantly,

Jojoba, Goatnut

used them also to make a drink, first grinding the nuts and then boiling the meal and straining off the liquid.

At one time the nuts were sold in Los Angeles drug stores as a hair restorer. To prepare this, they were boiled and the released oil was rubbed into the scalp or the eyebrows. This product seems to have had a wide reputation in the Southwest, and is still used in parts of Mexico.

The Mexicans made a rich drink of the Jojoba nuts, which they first roasted, then ground together with the yolk of a hard-boiled egg. The pasty mass was boiled with water, milk, and sugar. To improve the flavor a vanilla bean was added. This made a drink something like thick chocolate.

Beavertail (*Opuntia basilaris*)

To the Indians of the Southwest this was one of the most important of many Cactus species (see pl. 1). It is common on dry benches and fans over the Mojave and Colorado deserts, through to Utah, Arizona, and Sonora.

The young fruit was gathered in quantities in the

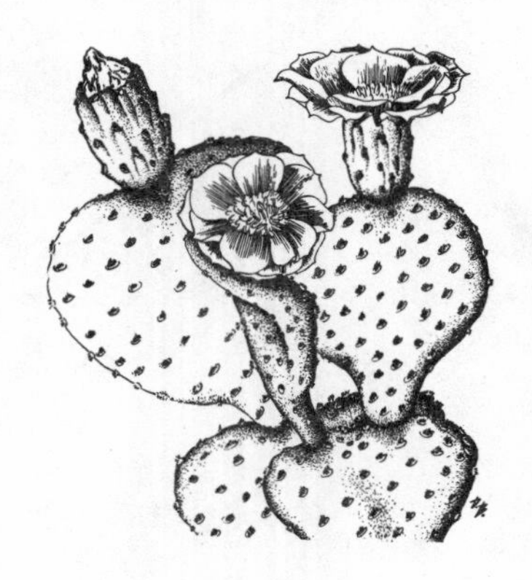

Beavertail

early summer. Easily broken off with a stick, it was collected into baskets. The fine short spines were brushed off with grass or a bunch of twigs. Cleaned, these young fruits were cooked in the stone-lined pit for about twelve hours (see p. 6). In the Panamint Mountains the Indians used not only the young fruits but also the flower buds and the young fleshy joints. With their spines removed by brushing, they were dried in the sun, a tedious process, and in that condition would keep indefinitely. As wanted, they were boiled and eaten with salt.

The pulp of older pads was scraped out and used as a wet dressing on cuts and wounds, changed frequently, and was said to deaden pain and to help in healing. The fine fuzz of spines was rubbed in to remove warts and moles.

Tuna, Indian-Fig (*Opuntia Fiscus-indica*)

This large-growing Cactus was introduced into California from Mexico and was much cultivated in hedgerows in the days of the Spaniards and the early settlers. Both Indians and whites made much use of the fruits of this plant and of several others which are very much like it, notably Prickley-Pear, *Opuntia occidentalis* (see pl. 4). The fruits were eaten raw, having been peeled carefully after removing the spines. A good syrup was made by boiling the peeled fruits and straining out the seeds. The Spanish settlers further reduced the syrup to make a paste, dark red or nearly black, known as *Queso de Tuna*. The flavor of the fruits of different species varies somewhat.

Another use for these plants, which is still current among the Mexican population, is to gather the young joints before the spines have hardened, cut them into strips and boil tender to serve as a vegetable. A good pickle is also made from them. The young growths are known as nopales and are considered a very fine vegetable.

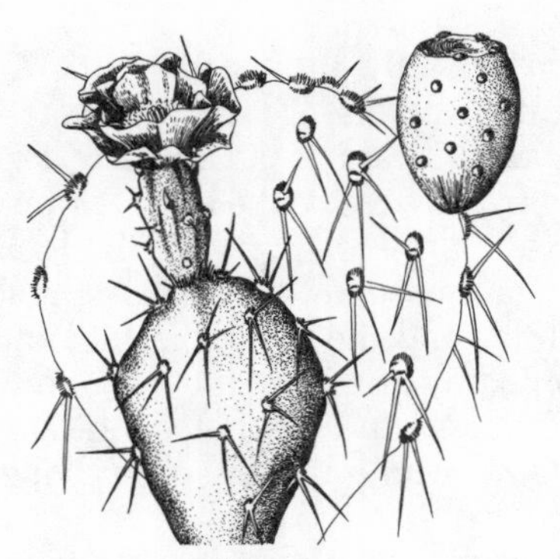

Prickley-Pear

Among the Indians the fully ripe fruits were gathered, dried, and the seeds winnowed out and stored, to be ground into flour for atole (see p. 6). The dry seeds of the Beavertail and several other Cacti, the ripe fruit of which is too dry to eat fresh like the Tunas, were used in this same way among the desert tribes. The mature fleshy pads of the Tuna were split, soaked, and used to poultice bruises. The pads were boiled and crushed, and the sticky juice resulting was added to mortar or used in whitewash to make it stick more securely to adobe walls.

Toyon (*Heteromeles arbutifolia*)

Known also as California-Holly and Christmas Berry, the Toyon is a common shrub or small tree on brushy slopes and in canyons throughout much of California and into northern Lower California. The bright scarlet berries were gathered by many of the Indian tribes. Rarely eaten raw, they were cooked either by roasting over hot coals, the bunches of berries being held over the fire, as they were gathered, or by tossing in a cooking-basket with hot pebbles or wood coals. This slight cooking seems to take away the somewhat bitter taste of the fresh fruit.

Spanish Californians also used these fruits, cooking them by putting them into boiling water, or even boiling them slightly and straining off the water. The berries were then wrapped in a hot cloth and left to steam for as long as two hours, after which they were ready to serve. Another way was to put the fruit into a bag, sprinkle with sugar, and place, covered, in a slow oven "for a while." A pleasant "cider" was also made from these berries by Spanish Californians and American settlers.

The Indians in some areas made a tea from the bark and the leaves which was used as a cure for stomachache and other aches and pains.

The fishermen of Catalina Island are said to have used the bark of this tree in tanning their nets and sails.

The name Hollywood was given on account of the number of California-Hollies which grew in the hills about the original subdivision.

Toyon

Manzanita (*Arctostaphylos*)

Over the greater part of montane California there is often a heavy covering of Manzanita (see pl. 6). Most of the forty-three species in the state were used for food in the early days. The berries were gathered green by the Spanish settlers to make a soft drink or a jelly. The Indians usually collected only the ripe fruit, beating it into their collecting baskets as they did the seeds of grasses. Several of the Indian tribes celebrated the ripening of the Manzanita harvest with a big feast and dance.

Often the berries were eaten fresh. Large quantities were dried and stored for winter use. The pulp, which is dry and sweetish when ripe, was easily reduced to a fine powder and separated from the seeds and skins to make a favorite drink. Mixed with water it was allowed

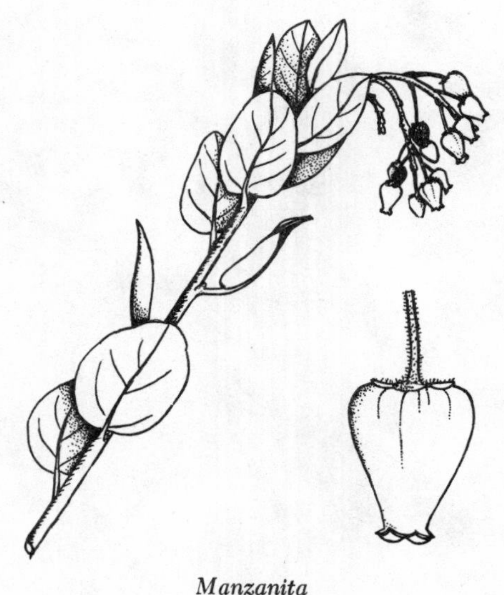

Manzanita

to stand for a few hours before use. In other tribes the whole berries, slightly crushed, were put into water and left to stand. The seeds and pulp were then strained out through a basket-strainer.

The dry, rather wooden seeds were beaten to a fine flour and made into mush or shaped into thin cakes which were baked in hot ashes. At times the meal was eaten dry, in pinches, like pinole (p. 6). The green fruit is very acid but is said to be good to quench thirst.

The leaves of two or three species are mentioned as being mixed with smoking tobacco. This was probably only done after the arrival of the white settlers. It is not at all clear that it was done earlier.

A wash or lotion made from the leaves was used as a cure for the inflammation caused by Poison-Oak, *Rhus diversiloba*. A tonic made from leaves was used to cure severe colds, but was too strong to be taken internally. Boiled down to a yellowish-brown extract it was used as a wash for body and head, to stop certain types of headache. The Concow Indian women chewed the leaves into a thick pad and put the mass as a poultice on sores.

The wood was used in making the Indian huts. Two V-forked sticks were selected for carrying bundled wood or fuel. The Karok Indians made spoons from this wood (this was undoubtedly after the arrival of the white settlers), sticks for scraping Acorn soup off the sides of the cooking-baskets, reels for string, and tobacco pipes. The wood pipes were sometimes soaked in eel or bear grease to prevent splitting, having first been boiled in water with the heartwood removed.

Mexican Tea, Squaw Tea (*Ephedra*)

Of the seven species of Mexican Tea which are to be found in California, only one grows west of the Sierra Nevada. They all seem to have been used, without distinction, for the same purposes. The names Desert Tea and Mormon Tea are almost as common for these plants as Mexican Tea. The Indians, Mexicans, and Spanish settlers all brewed a pleasant refreshing drink by steeping the stems, either green or dried, in boiling water. The length of time in brewing depended on the strength of tea required.

This tea was also used as a tonic for kidney ailments, to purify the blood, and for colds, stomach disorders, and ulcers. The dried stems were ground to powder and used on open sores or mixed with the resin from the Pinyon Pine and used as a salve. For burns the powder was slightly moistened and used as a poultice.

In a number of areas, the local *Ephedra* was used in combination with other plants: Cushion Gilia, *Ipomopsis congesta;* Scarlet Gilia, *Ipomopsis aggregata;* Antelope Bush, *Purshia tridentata;* or the scraped bark

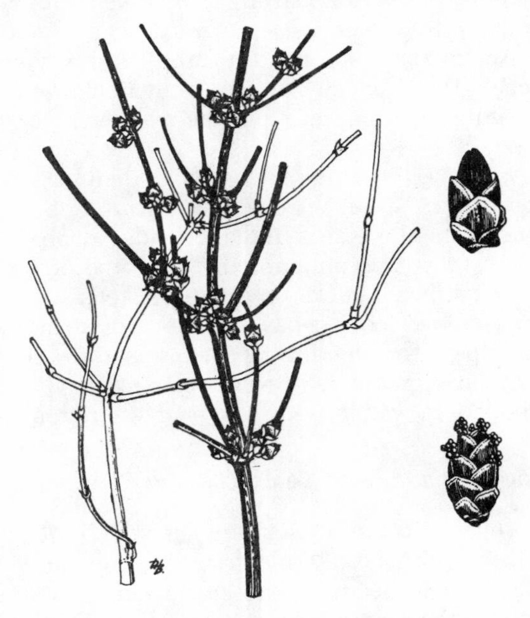

Mexican Tea, Squaw Tea

of Mountain-Mahogany, *Cercocarpus ledifolius.* An early record tells that the Panamint Indians roasted and ground the seeds of *Ephedra nevadensis* to make a bitter bread.

Barberry, Mahonia (*Berberis*)

In most of the mountainous parts of California there are Barberries (see pl. 4). Of the thirteen species now recognized scattered through the country, seven occur in the records of the early uses of plants. The fruits of most were used by both the Indians and the early settlers to make a pleasantly acid drink. These berries were also used in making a tart preserve, and some of the Indian tribes dried the fruits for winter use. Probably the most important use for these plants was in making a good yellow dye for baskets, buckskins, and fabrics.

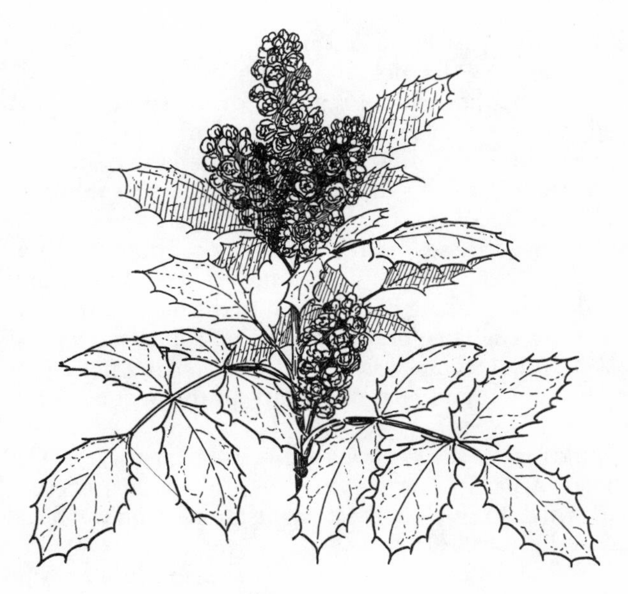

Barberry, Mahonia

The roots, and in some cases the stems also, were crushed and boiled to obtain the color.

The bark of both *Berberis repens* and *Berberis pinnata* was used as a laxative and to make a lotion to treat various skin diseases. From the roots a bitter tonic was made which served as a blood purifier. The leaves of *B. repens* were boiled and the tea taken to cure general aches and rheumatic pains.

The Karok Indians considered the blue berries of Oregon-Grape, *Berberis aquifolium,* to be poisonous, but pounded them, fresh, with the flowers of Larkspur, *Delphinium decorum,* and salmon glue to make a paint for decorating bows and arrows. These berries, far from being poisonous, were, and perhaps still are, made into a very good jelly by the Oregonians.

Barrel Cactus (*Echinocactus acanthodes*)

Although popularly looked upon as a source of water in the desert, it is not always understood that these plants do not hold water ready to drink. In case of real need, one has to slice the top off the barrel, then pound the flesh until the liquid is squeezed from the pulp. This is not the refreshing drink one would like to think of, but it will save one from dying of thirst. Some of the Indians used these "barrels" for cooking pots. The top was sliced off and the flesh scooped out. As cooking was done by placing hot stones into the meal in the pot, this was as satisfactory a cooking vessel as the woven basket. When camp was moved, the cooking pot could be left behind and new ones made at the next camping ground.

Spines from these plants were used for awls, set in a head of hard pitch. They were also used as needles with which to prick in the designs tattooed on the faces and bodies of the Indians.

The flesh of another of these barrels, known as Niggerheads, *Echinocactus polycephalus,* was made into a

Barrel Cactus

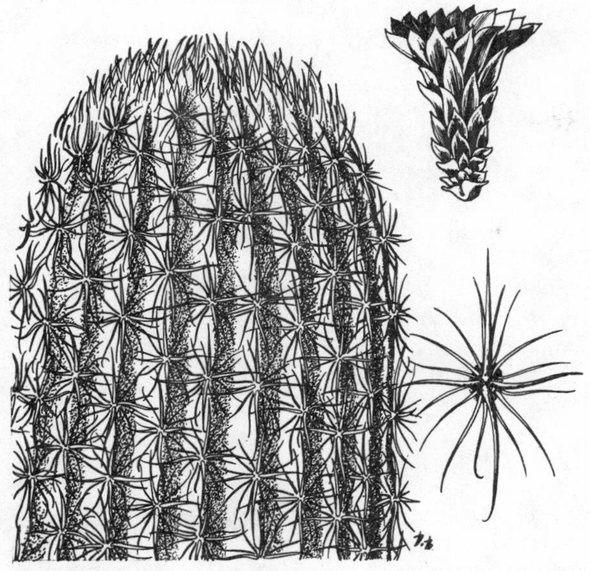

kind of candy by the Papago Indians. The thorny rind was cut away and the remainder left to drain for a few days. The pulp was then cut into pieces which were boiled in syrup made from the fruit of the Sahuaro, *Cereus giganteus*. This type of cactus candy is still made, though most generally with sugar rather than Sahuaro syrup.

Sugar Bush (*Rhus integrifolia*)

Two shrubs common in the southern part of the state, more particularly toward the coast, were used in the making of popular summer drinks. The ripe berries of the Sugar Bush, *Rhus ovata* (see pl. 4), and the Lemonade Berry, *Rhus integrifolia*, are coated with a sour-sweet, sticky substance which both the Indians and the early settlers enjoyed when stirred into water and cooled.

Spanish Bayonet (*Yucca baccata, Yucca schidigera*)

Throughout the mountainous areas of the Southwestern deserts there grow several kinds of Yuccas, commonly called Spanish Bayonet or sometimes Amole. The Indians of southern California used two, *Yucca baccata* and *Y. schidigera,* though the large fleshy fruits of *Y. baccata* (see pl. 4) were prized above all others for food.

The Spanish Bayonet was by far the most important plant in the Southwest for the production of fiber, apart from the many other uses to which the Indians put it. All parts of the plant were used. The very earliest rec-

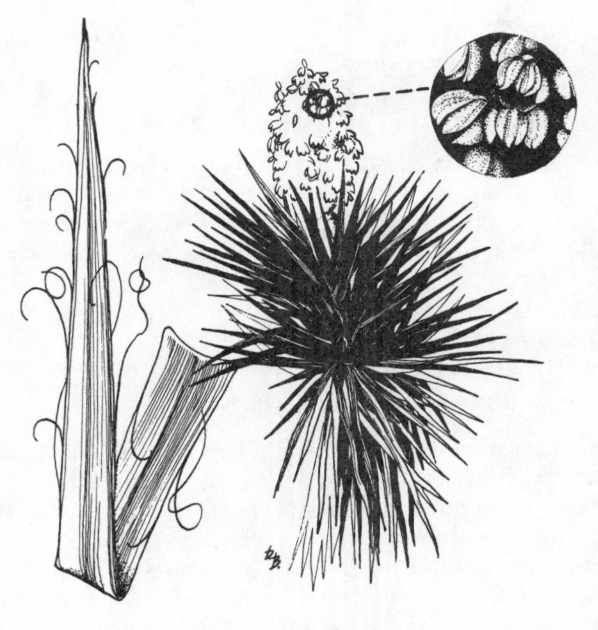

Spanish Bayonet

ords show that prehistoric Indians made sandals, cords, baskets, and rough kinds of cloth from these fibers. Some of the sandals were made of whole leaves plaited, some with split leaves, and in others the fibers were freed from the fleshy parts of the leaves, made into cord, and plaited or woven.

To obtain the fibers the green leaves were first soaked in water, then pounded on a flat rock with a wooden mallet or beater, and plunged into water from time to time during the process to wash out the skin and the softer tissues. The fibers so released were fine, strong, and white. Cords and threads of different sizes were made for tying, binding, and sewing.

For tying together the poles of the framework of their houses, whole leaves were often used, and the grass covering or matting used on the walls was sewn onto the framework with Yucca thread.

A list of articles made by the Indians from Yucca fiber (compiled by Dr. Edward Palmer in 1878) includes "ropes, twine, nets, hats, hairbrushes, shoes, and mattresses." He also describes excellent horse blankets made from these fibers and woven by the Diegueño Indians on a primitive loom of native origin.

Pottery making was known only to a few of the desert and river tribes of the Southwestern states, so that most of the Indians were dependent on baskets for their cooking, storage, and carrying vessels. A red strand for making patterns in basket work which was obtained from the inner bark of the roots of the Joshua Tree, *Yucca brevifolia* (see pl. 5), was the most highly thought of for this use, though some was also obtained from the Spanish Bayonets. Also a brown fiber for these patterns was obtained from the Joshua Tree by some of the Colorado River Indians. A finer whiter fiber was extracted from the leaves of Our Lord's Candle, *Yucca Whipplei* (see pl. 5), from which good thread was manufactured.

The brushes used for painting the designs on pottery

were made from Yucca leaves cut to the required length and width and bruised at the end to make a fringe.

The fruit of the Spanish Bayonet was a very important item in the food supply of many of the tribes of the Southwest. As the deer, birds, and insects also fed freely on this fruit, it was often gathered while still green and ripened in the sun or in the houses. It was sometimes cooked green and eaten immediately. The ripened fruit was eaten raw or much more generally cooked, either by boiling or by pit-roasting (see p. 6). After the skin and the seeds had been discarded and the fruit cooked, it was worked into a form of paste which was spread on basketwork mats to dry, sometimes for several weeks. The paste would be worked and reworked into blocks or cakes which could be stored for winter use or for trading.

The fruits were also split and dried in the sun after the seeds had been removed. The dried product could be eaten as it was or boiled up with water. Frequently pieces were broken off, dissolved in water, either hot or cold, and used as a sweet drink. While the extent to which the different tribes made use of this food varied considerably, the methods of preparing the preserved product seem to have been rather uniform. For those in whose territory the plants grew, the dried product was an important article of barter with the neighboring tribes.

There seems to be no record that the seeds of the Spanish Bayonet were used in any way, though the seeds of both the Joshua Tree and Our Lord's Candle were very largely used, being ground to flour and eaten dry or made into mush. The young flowers, flowering stems, and buds of several kinds of Yucca were gathered, cooked in the stone-lined pits, and eaten, though the large flowers of *Yucca baccata* were said to be bitter if gathered after the summer rains.

The dense young flower heads of the Joshua Tree were pit-roasted by the Indians of Death Valley. To secure the head the stiff leaves below the flowers were grasped, pulled over the head, and snapped off with a sidewise twist to the end of the branch. The roots and stems of both species of Spanish Bayonet were sometimes used in the place of soap. The large fleshy roots were washed clean, pounded on a stone, and then rubbed into the scalp or on the clothing to be washed. This form of soap was highly prized both by the Indians and the early settlers. The roots could be dried and kept for long periods before being used. Sometimes a washing fluid was made by boiling the roots. There are records that the Yucca leaves were sometimes used, though the soap in them is not so abundant as in the roots.

Bear-Grass (*Xerophyllum tenax*)

Abundant in the mountains parts of northern California, below about 6000 feet, and across into Montana and

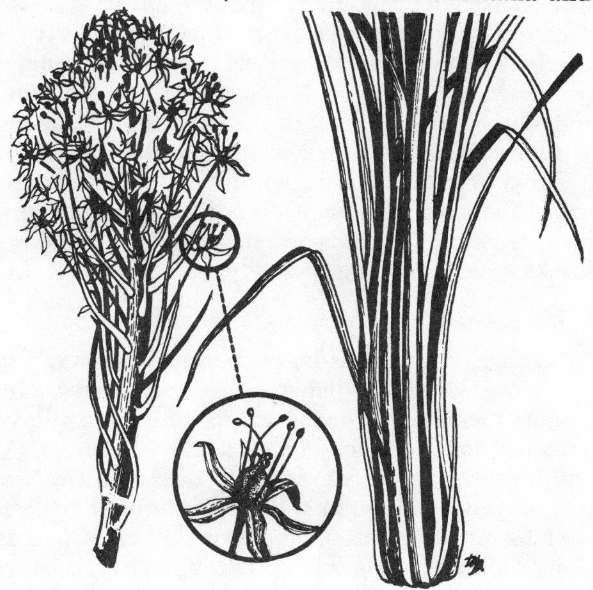

Bear-Grass

northward into British Columbia, this plant has surely captured the popular fancy from the number of "common" names given it: Squaw-Grass, Elk-Grass, Fire-Lily, Turkey-Beard, Bear-Lily, Pine-Lily.

Fibers split from the leaves were used in much of the finest basket work of many of the Indian tribes of central and northern California. These strands were used generally for the white overlay in woven baskets, sometimes as patterns, with the black strands of the Five-Finger Fern, *Adiantum pedatum*, and the red-dyed root fibers of the Chain Fern, *Woodwardia fimbriata.* Very fine baskets, women's caps, and watertight cooking vessels were sometimes completely overlaid with this beautiful white fiber.

Some of the Indians insisted that the area from which the Bear-Grass was to be gathered must be burned over in the preceding year so that only the new green leaves would be there. These were gathered in June and July. The young growth was selected as it was more easily handled and worked. The fibers were sometimes split from the leaf, using the woman's hair, held taut, inserted in the end of the leaf, the latter then being pulled so that the hair split the leaf full length.

Dress ornaments and some fabrics were made from these fibers, though these do not appear to have been at all general. There is reference to the bulbous root stocks as a nourishing food when roasted.

Deer Grass (*Muhlenbergia ringens*)

In most of the records on early plants, this grass is referred to under the name *Epicampes ringens.* It is a rather wiry grass growing up to three feet tall, widespread, though no longer abundant in southern California, spreading to the foothills of the Sierra Nevada and to New Mexico and Mexico. It was used extensively by most of the tribes of southern California in making the foundation of their coiled basketwork. A small bunch of the grass was bound with a strand of Rush or

Islay, Holly-Leaved Cherry

Prince's Plume

Mariposa-Lily

Beavertail

PLATE 1

California Buckeye

Mescal, Agave

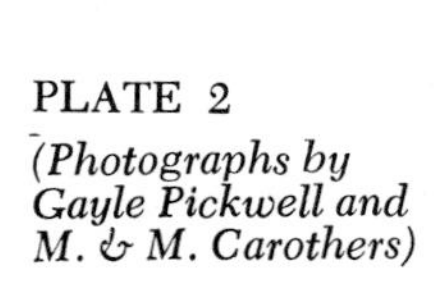

PLATE 2
(Photographs by Gayle Pickwell and M. & M. Carothers)

Screwbean
(Photographs by Gayle Pickwell)

PLATE 3

Pinyon
(Photographs by Ernest S. Booth)

Prickley-Pear

Barberry, Oregon-Grape

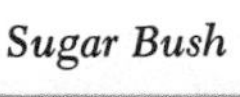

Sugar Bush

Spanish Bayonet

PLATE 4

Joshua Tree

Our Lord's Candle

Iris

Poison-Oak

PLATE 5

Manzanita

PLATE 6
(Photographs by Ernest S. Booth)

White Alder

Soap Plant, Amole

PLATE 7

Madrone
(Photographs by Pickwell, Booth, and M. & M. Carothers)

Squaw Bush
(Photograph by M. & M. Carothers)

Jimson Weed

Douglas-Fir

Milkweed

PLATE 8

Squaw Bush. This formed the coil which was stitched with the wrapping strand to the coil below, so shaping the article being made. Further north other materials were substituted, and the grass was not used at all by the tribes whose baskets were woven or plaited. There is a story of a group of Indian women who regularly traveled a distance of some twenty miles to a particular canyon to gather supplies of this grass, which must be taken at exactly the right time, neither too green nor over-ripe. Those Indians who used it used this species exclusively, refusing any others which might look something like it. It is interesting that there seems no record of any other grass having been used for this purpose.

Indian-Hemp

(*Apocynum cannabinum* var. *glaberrimum*)

This well-known source of fiber is to be found occasion-

Indian-hemp

ally in damp places, below 5000 feet altitude through most of California, even here and there on the deserts. The species extends into Canada and to the Atlantic Coast. The work of preparing the fiber from the stems of this plant was the same as with that obtained from the Spanish Bayonet and Mescal plants. The fiber was used particularly in making fishing and carrying nets, for string and for ropes, and to some extent for weaving rough cloth. In the early days of American settlement it is recorded that the Swedes of the Delaware River colonies bought ropes of this fiber from their Indian neighbors "at the rate of fourteen yards for a piece of bread," preferring it to ropes made from the common hemp which were more easily available to them. (Unfortunately, the size of the "piece of bread" is not given.)

Iris (*Iris Douglasiana, Iris macrosiphon*)

Iris Douglasiana (see pl. 5) grows abundantly along the coast from north Santa Barbara County into Oregon. *Iris macrosiphon* does not occur on the west slopes of the outer Coast Ranges but is to be found over a very large area throughout the rest of those ranges from Santa Cruz County northward. From its distribution in the North Coast Ranges, it would seem very probable that *Iris Purdyi* should also be included with the other two as a source of fiber for the Indians.

The Iris leaves were gathered in large bundles and a single silky fiber was taken from each margin of the leaf. None of the other fibers was used. Using a mussel-shell or abalone "thumbnail" the women cleaned these fibers. The sharpened shell was either attached to the thumb or held in the palm of the hand. With this instrument the fibers were detached from the leaves and scraped clean of all tissues. The threads were twisted on the bare thigh by the men, mostly while sitting around in their sweathouses. The men always knotted the fishing nets.

This fiber makes a beautiful strong and pliable cord or rope. One Indian stated that "it takes nearly six weeks to make a rope twelve feet long." In spite of the tremendous labor of preparing this material, the Iris fiber was one of the most generally employed in northwestern California. From the threads and cords made of this fiber the Indians made their fishing nets and camping bags as well as snares for catching deer, birds, and other game.

It is said that long ago the Yokia squaws of Mendocino County wrapped their babies in the soft green leaves of *I. Douglasiana* while on the hot dry hillsides collecting Manzanita berries. This wrapping retarded perspiration and saved the babies from extreme thirst.

Poison-Oak

(*Rhus diversiloba, Toxicodendron diversilobum*)

Everyone should know Poison-Oak (see pl. 5). It is common in thickets and on wooded slopes through a great part of California west of the Sierra Nevada. It is to be found in the foothills, along the mountain streams, and in washes and hedgerows, below 5000 feet. In spite of its poison, this plant was used in a number of ways by many of the tribes in the areas where it grows.

More particularly among the northern tribes, the supple slender stems were used for the warp or ground in woven baskets. The leaves were used by the Karok Indians of northwestern California to cover the bulbs of Soap Root, *Chlorogalum pomeridianum*, while baking them in the earth oven for food (see p. 6). The Concow Indians of Marysville Buttes mixed the leaves into their acorn meal when baking bread, and other tribes used the leaves to wrap the meal during the baking. Poison-Oak twigs were used to spit salmon steaks while they were being smoked.

The juice from stems, leaves, or roots was used as a cure for warts. The wart was either cut off close to the base or pricked or cut several times, and the fresh juice was applied immediately. The application had to be repeated several times before the wart would disappear. This juice was also used as a cure for ringworm. Fresh leaves bound tightly over the bite of a rattlesnake were thought to counteract the poison, but this had to be done immediately. The fresh juice turns black quickly and was used in making an excellent black dye for basket materials. Some of the purest black strands in the baskets made by the Pomo Indians were dyed with this juice.

The full-blooded Indians seem to have been only slightly, or not at all, subject to the effects of this poison, though those with mixed blood often suffered as badly as their white fellows. Some are said to have eaten a

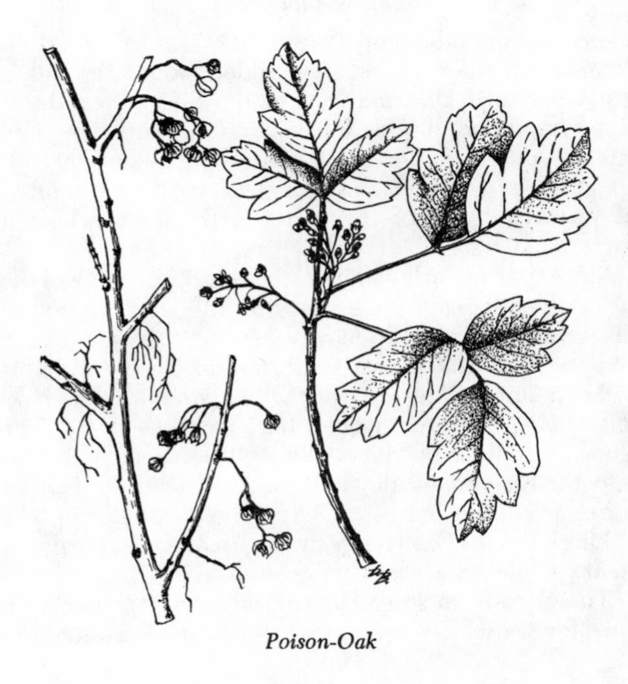

Poison-Oak

small piece of the very young leaf in the early spring to prevent the poison from affecting them for the rest of the year. It is clear that many could not have been subject to the effects of this poison or the plant could not have had so many uses in their everyday lives. For remedies in cases of poisoning the Indians were taught by the early settlers to use such salts as baking-powder. Possibly the use of Soap Root, *Chlorogalum pomeridianum,* cooked and made into a paste was also learned from the white man. One of their original cures seems to have been a strong extract from the root of the Sunflower, *Wyethia longicaulis,* with which the inflammation was bathed.

Squaw Bush (*Rhus trilobata* var. *malacophylla*)

Common in the canyons and washes of the interior valleys of California, mostly below 3500 feet, Squaw Bush extends northward into Oregon (see pl. 8). The desert form, var. *anisophylla,* spreads eastward into Utah and Arizona. In the south the tough, slender branches were stripped of their bark and carefully split into several strands which were used to wrap round the foundation of Deer Grass in making coiled baskets. Some of the northern tribes employed the withes of the *Rhus* as a foundation on which to weave, but Squaw Bush was much more extensively used in the south than in the north. The strands used by the southern Indians were sometimes blackened by soaking them in a dye made from the berry stems of the Elderberry, *Sambucus mexicana,* for a week or more. Undyed, the strands were a light straw color. The deep black strands were used in making the patterns.

The Luiseño Indian women made a fan or beater of these twigs with which to beat the small dry seeds of grasses and annuals into gathering baskets.

The fruits which in the desert variety are a bright red were used in making a slightly acid, refreshing drink by soaking them in water. In some tribes the berries were dried, powerded, and used for food. They were also employed in the treatment of smallpox. Dried and powdered, they were dusted on open sores; the powder was mixed with water for a lotion used to bathe the dry rash. A decoction made from the stem was said to be a good cure for coughs and lung ailments.

Yerba Santa (*Eriodictyon californicum*)

Several species of *Eriodictyon* are known by this name. *E. californicum* grows to about 4000 feet on dry mountain slopes and ridges through the coastal ranges and up into the foothills of the Sierra Nevada from Monterey and Tulare counties northward. *E. trichocalyx* is the southern counterpart in drier places up to 8000 feet from Ventura County southward. *E. angustifolium* is in the mountains of the eastern Mojave Desert, spreading into Arizona and Utah.

To a large extent these three plants were used for the same purposes. The name "Yerba Santa" (Holy Weed) was given by the Spanish Fathers who early learned from the native Indians the medicinal value of these shrubs. The thick, sticky leaves, either fresh or dried, were boiled to make a bitter tea, taken as a cure for

Yerba Santa

coughs, colds, sore throat, catarrh, and asthma, also for tuberculosis and rheumatism. It was thought to be a blood purifier and was taken frequently, in a weaker form, in the place of regular tea. A liniment was used as a wash to reduce fever. The fresh leaves, pounded into a poultice, were bound on the sores of both men and animals. A very strong solution from the boiled leaves eased sore and fatigued limbs. The young leaves and stems were used as a hot compress to cure rheumatism.

In some areas the leaf served as tobacco both for smoking and chewing. The fresh leaves are also said to be wonderful to quench thirst, just chewed as one goes along the trails. The first bitter taste shortly disappears and is replaced by a sweet, cooling sensation.

Yerba Mansa (*Anemopsis californica*)

Common in wet, alkaline places through southern California, the Central Valley, to Lower California, Ne-

Yerba Mansa

vada, Texas, and Mexico, Yerba Mansa was early adopted by the Spanish settlers. The peppery aromatic root was dried, then chewed raw for ills of the mucous membrane, or applied as a powder to heal knife-cuts.

More generally a tea or wash was brewed from the root and used as a liniment for skin diseases, cuts, bruises, or sores for both man and animals. The tea was also taken for indigestion, asthma, and to purify the blood. Reportedly one of the most popular remedies among the Mexicans, Yerba Mansa was thought to be almost a cure-all and was used most extensively. By heating the leaves, a wilted poultice was made and used to reduce swellings. A bath to relieve muscular pains and sore feet was made by boiling the leaves in a quantity of water and soaking the aching part.

White Alder (*Alnus rhombifolia*)

The White Alder (see pl. 6) is a common tree along the rivers and streams through a large part of California. It is not usually found above 5000 feet altitude.

A tea made from the bitter bark, either fresh or dried, was used in large quantities to produce perspiration and as a blood purifier. It was also used to cure stomachache and sometimes to check diarrhea caused by drinking bad water. It was also employed in specified doses as a cure for tuberculosis, by checking hemorrhage.

A red dye was obtained from this bark. The Indians chewed the fresh bark and used the colored spittle to dye the stems of the Chain Fern for making patterns in their baskets. The Wailaki Indians used this red spittle to color their bodies to help them in catching the red-fleshed salmon. The dye was dried on and then had some resistance to water. The naked Indians then went into the river to drive the fish into nets stretched across the stream. The coloring was to ensure a good catch.

Sometimes young Alder shoots were used to make arrows. The dry rotted wood was used with powdered willow-bark as a poultice on burns. The soft wood was considered valuable tinder. The roots of both White and Red Alder, *Alnus oregona,* were used in making caps and trinket baskets.

Jimson Weed (*Datura meteloides*)

This beautiful, poisonous plant (see pl. 8) grows in dry waste places over much of southern California and up into the lower Sacramento Valley. The common name, Jimson Weed, is said to be a corruption of Jamestown Weed, a name given in the Eastern states to an introduced relative, *Datura Stramonium.* Throughout southern California the plant was commonly known by the Mexican name Toloache, which comes directly from the Aztec.

Jimson Weed

Toloache had a number of uses among the Indian tribes, though it was always employed with great caution because of its poisonous qualities. Most frequently a liquid was brewed from the crushed root. Sometimes the seeds were crushed, soaked in water, and the mixture left in the sun to ferment. The resulting brew had the same narcotic effect as the root preparation, together with the added effect of alcohol. The dreams induced by this drug were usually the reason for using it. The crushed plant was sometimes used to bind on bruises and swellings, and there is record of its use in extreme cases of saddle sores on horses, and also as a cure for rattlesnake and tarantula bites. The leaves were dried and smoked as a cure for asthma, if the patient had a sound heart.

By far the most widespread use, however, was in the rites performed by boys entering manhood. There was considerable variation between the different tribes in the details of the performance, but the central act, the drinking of Toloache, the dance, and the importance of the dreams which the boys experienced afterward were universal.

In a much more mild form Toloache seems to have been used among some of the southern tribes in the rites for young girls preparing for womanhood, though the references do not show its use as so highly ceremonial, or so widely practiced.

Douglas-Fir (*Pseudotsuga Menziesii*)

Abundant throughout the Coast Ranges of central and northern California and along the western slopes of the Sierra Nevada from Fresno County northward, mostly below 5000 feet, this handsome tree (see pl. 8) was used in a number of ways by the Indians. It is now the most important lumber tree of North America.

Most of the information available comes from the tribes of the north coastal region of California. When a convenient fallen log was found, planks were made for the building of houses. The shaft of the salmon harpoon was made from this wood, being tipped with fore-shafts of a Service-Berry, *Amelanchier pallida.* The poles used with the "dip-nets" in fishing were also of Douglas-Fir, as were hooks used as an aid in climbing the Sugar Pine to collect the seed for food.

At least one tribe used the smaller roots in the making of baskets. Roots about the size of a pencil, often eight to ten feet long, were dug, split into numerous strands, and woven about a framework of hazel or willow withes. The roots of both Yellow and Jeffrey Pines, *Pinus ponderosa* and *P. Jeffreyi,* were also used for basket work, and were gathered after the trees had flowered so that they were tough enough and extended beyond the spread of the lower branches. The roots were cleaned, placed parallel in a heated pit, and covered with earth; a fire was built on top and kept going for two days. When the roots were light in color they were ready to be split into four pieces. Those which were to be used at once were split into smaller strips with a deer bone awl until reduced to the size required. They were scraped with a mussel shell to make them more soft and pliable. The remainder were tied in bundles and put away until needed. *P. Jeffreyi* was used especially in the making of tobacco baskets, *P. ponderosa* in making cups and acorn baskets. A hot drink, said to be used as a substitute for coffee by both Indians and whites, was made from the fresh young needles of the Douglas-Fir. A similar brew of the tips of the branches was used medicinally for tuberculosis and lung troubles. As a cure for rheumatism, branches of Douglas-Fir were laid on a bed of hot rocks covered with earth, in the sweathouse, or temescal. A blanket was laid over the branches and the patient lay on top. He may have been there as much as half a day at a time while steam from the earth came up through the boughs. Branches were also waved over the patient while the shaman chanted charms.

As a charm, a hunter's bow and arrows would be passed through the smoke from Douglas-Fir boughs placed on the fire while a "formula" (incantation) was sung. This was supposed to keep the deer from scenting the hunter and to secure success to the hunt.

Soot gathered from burning the gum in a hollow tree was rubbed into the punctures of the three parallel lines of the tattoo marking a girl's chin.

Madrone (*Arbutus Menziesii*)

This very beautiful tree (see pl. 7) of the west coast, scattered in the south from Lower California to San Luis Obispo County and becoming abundant from there northward to British Columbia, had many uses among both the Indians and the early settlers. An infusion brewed from the root, bark, or leaves was used as a treatment for colds. One tribe used the bark, which peels and falls off every year, to make a tea to cure stomachache. Both leaves and bark went into the making of a lotion to bathe sores and cuts, which were said to heal quickly under this treatment. This same cure was used in treating sores on horses.

The handsome, scarlet berries were eaten by both Indians and early settlers. They were eaten fresh, as gathered from the trees, or more frequently they were cooked in the Indian fashion with hot stones in water in a cooking basket. The berries were then dried and could be stored, to be soaked in warm water when wanted, or mixed with crushed or ground Manzanita berries.

The Spanish settlers made stirrups from the hard wood of the Madrone, and some of the Indians used the straight, slender trunks for lodgepoles in their dwellings and fashioned various small tools, like digging sticks, of this wood. Madrone charcoal was employed by early Californians in the making of gunpowder; one reference states that it was highly thought of commercially.

Death-Camas (*Zigadenus venenosus*)

Almost the only native bulbs which were not used for food by the Indians are the Zygadenes. The Death-Camas (one of the Zygadenes) grows in moist grassy places in much the same area as the Camas and was occasionally taken by mistake with fatal results. The flowers are a cream-white and not so large as the blue flowers of the Camas. The crushed bulbs were used as poultices for the cure of boils, bruises, strains, rheumatism, and in some areas for rattlesnake bites. Occasionally the bulbs were roasted and applied warm. On the desert side of the Sierra Nevada the Death-Camas is replaced by a smaller plant, *Zigadenus paniculatus*, often called Sand-Corn, from the number of small bulbils which form around the adult bulb. This was used in the same way and for the same ailments as the Death-Camas. An emetic tea was also made by boiling either of these bulbs.

Death-Camas

SOAP AND FISH POISON PLANTS

Amole (*Chlorogalum pomeridianum*)

This Amole or Soap Plant is to be found on dry slopes and mesas, often in washes and pastureland, west of the Sierra Nevada to an altitude of about 5000 feet, from southern California to southern Oregon (see pl. 7).

The primary use made of this bulb by the Indians and later by the early settlers, both Spanish and American was, as its common names show, in the place of soap. The bulb, stripped of its harsh, fibrous outer coating, was crushed and rubbed on hands or clothes in

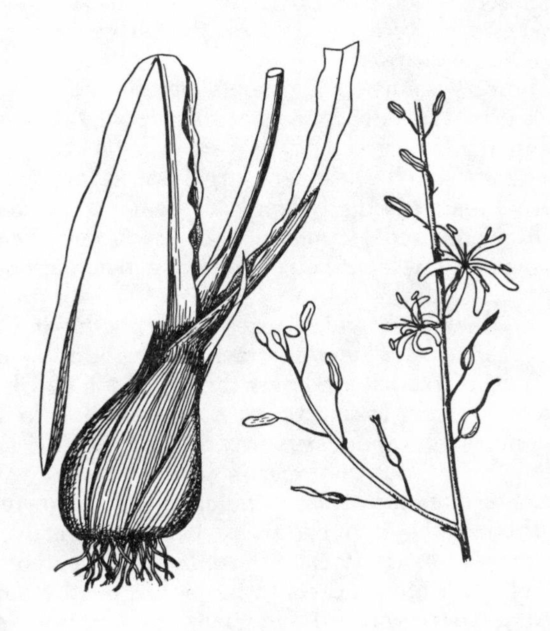

Amole

water to make an excellent lather. Wherever the plant grew, it was used for all kinds of washing jobs. It was considered to be an excellent shampoo, leaving the hair soft and glossy, and was said to be useful in removing dandruff. The absence of alkali makes it especially good for washing delicate fabrics. The bulb can be dried and stored for use as wanted. Some of the Indians used it for food, baking the bulbs in the stone-lined pit (see p. 6). The bulbs put into the oven at night were ready in the morning. The slow cooking destroys the soapy characters leaving a good nourishing food. The cooked bulbs contain a considerable amount of fiber which the Karok Indians saved and made into small brushes for sweeping the fine flour out of the inside of the basket-hopper in the process of grinding acorn flour.

The coarse outside fibers were also used in making brushes, and were gathered for the early white settlers for filling mattresses.

The very young spring shoots were very sweet when cooked by the slow process of the pit-oven. While still young, the fresh green leaves were also sometimes eaten raw, and older leaves were largely used for wrapping acorn bread during baking. Medicinally the roasted bulbs were used as poultices for sores, and the fresh crushed bulb was rubbed on the body to cure rheumatic pains and cramps.

The thick juice which oozes out of the baking bulbs was used as a glue for attaching feathers to arrow-shafts and was smeared over the wood of a new bow to take a covering of soot to make the new bow look old. The juice from the leaves was pricked into the skin to make green tattoo markings.

Large numbers of fish were caught by throwing the crushed bulbs into carefully dammed streams. The effect was to stupefy the fish so that they floated to the top of the water and could be picked out by hand or with a coarse-meshed net. This coöperative fishing, where the whole village worked together, meant that great quantities of bulbs would be prepared at one time. An interesting note on this procedure says, "All the fish and also the eels, but not the frogs, were stupefied." The poisoning of fish in this manner does not affect their value as food, and no ill effects have ever been reported from the eating of fish so caught. A number of other plants were also used for this purpose in California, but none quite so widely as the Amole. This way of catching fish is well known to the natives in other parts of the world, using other plant poisons. It has long been illegal in California because it destroys all the fish, both large and small, and rapidly empties the streams and lakes of fish.

Soap Plant (*Chenopodium californicum*)

A common weed in canyons and dry shady places below 5000 feet, it grows in the Sierra Nevada foothills and the Coast Ranges. The long fleshy roots were much used both by the Indians and the early settlers in the place of soap. The fresh tubers were ground on rock or crushed, and the mashed result was stirred or beaten in water to make a lather. The roots were also dried and stored; they became almost as hard as rock. When needed, they were grated or ground to a powder and used in the same way as modern soap powder.

Calabasillo (*Cucurbita foetidissima*)

This plant is common in sandy and gravelly places below 2000 feet through cismontane California from the San Joaquin Valley southward to San Diego County. East of the Sierra, it goes to 4000 feet across the Mojave Desert to Nebraska and Texas.

The large fleshy root was cut into pieces which could be used as cakes of soap or crushed and lathered in water. The green fruit, crushed with a little soap, was used for removing stains. The gourds, when ripe, were dried and put away for later use as the dried pulp was as effective as the fresh fruit.

The seeds are said to have been dried and made into

a mush for food. Leaves, fruit, and root were used medicinally. Pulp of the green fruit mixed with soap was put on sores and ulcers. Pieces of the root were roasted in hot ashes and rocks, thoroughly dried, and a small piece of the dried result boiled in water. One cup or less was the dose which acted as an emetic and a physic. Sometimes the dry powdered seed was dusted on open sores. It was claimed that the decoction from the root would kill maggots in wounds. The root crushed in water was applied as a dressing to piles.

A curious poultice for saddle sores on horses was made by mixing the crushed root with sugar.

Saltbush (*Atriplex californica*)

This plant grows along the coast from Marin County to Lower California. The long fleshy roots were used

Saltbush

as soap by Indians, Mexicans, and early settlers alike. The root was used in the same manner as the Soap Plant, *Chenopodium*, and the resulting suds were said to be particularly good for washing woolen fabrics.

The Indians also made good use of the seeds of this plant, gathering them in large quantities and using them to make mush or bread. The seed was cleaned and parched by tossing with hot coals in a basket. It was then ground and used in several ways. Sometimes the seed was ground without parching and made into the standard mush. This is one of the pinole seeds (see p. 6).

Turkey-Mullein (*Eremocarpus setigerus*)

One of the most abundant annual weeds over large areas of California, making most of its growth after the

Turkey-Mullein

summer rains, this low gray plant was used freely by both the Indians and the early Spanish settlers for catching fish. The whole plant was gathered and crushed and thrown into the pools and streams in the same way as Soap Plant was used. This use of the plant gave rise to the Spanish name for it, *Yerba del Pescado.*

Another name, Dove Weed, comes from the fact that the wild mourning doves feed greedily on the abundant seed. The Indians, taking advantage of this, would go to the areas where the Dove Weed was thick and kill large numbers of the feeding doves for food. Turkeys also are very fond of this seed, from which fact came the name Turkey-Mullein, the leaves of the plant being somewhat like Mullein, *Verbascum,* leaves.

The fresh leaves were bruised and applied as a counterirritant poultice for internal pain and asthma. The fresh leaves in warm water made a bath or wash used in typhoid and other fevers. A very weak solution was sometimes taken internally as a cure for chills and fevers. A poison for tipping arrows was also made from this plant.

DYE, GUM, AND TOBACCO PLANTS

Indigo Bush (*Dalea Emoryi*)

A small shrub, four or five feet tall, the Indigo Bush, is scattered in dry open places below 1000 feet through the Colorado Desert to Arizona and Lower California. The little bush is pungent and covered with strong yellow-orange glands which give off the dye used by the Indians.

Small branches steeped in water give a bright yellowish-brown dye which the Indians of the desert tribes used for coloring their deer skins and for dyeing strands of Rush and Squaw Bush used in making patterns in their baskets.

Indigo Bush

It is reported that the seed was used for food, but this cannot have been a very general practice as the statement does not occur commonly in the literature.

There is little reference to medicinal uses for Indigo Bush, but the closely related *Dalea polyadenia* was used not only for the dye it produced but also in a large number of ills and ailments. A tea made from the boiled stems, either fresh or dried, was usually taken hot as a cure for colds and coughs. In some tribes both leaves and flowers were used with the stems, but the stem decoction was thought of most highly for pneumonia, tuberculosis, and influenza. A cure for ills of the kidneys was made by boiling the tips of the shrub with chopped twigs of Utah Juniper, *Juniperus osteosperma*. The tea from *Dalea* was used much in cases of smallpox, given internally in small doses and externally as a cleansing wash. The stems were chewed as a cure for toothache and neuralgia, and the hot solution was used as a lotion for rheumatism. The dried stems were ground to powder and dusted on open sores. Medicinally this plant seems to have served the desert tribes for the same type of cure-all that the *Yerba Santa* served the more western areas.

Milkweed (*Asclepias*)

Almost all over California the Milkweeds (see pl. 8) can be found in waste ground, on rocky hillsides, or in the sandy deserts. Of the fourteen species now recognized, no more than five appear in the literature on early uses of plants. Though most of these plants are thought to be poisonous, some to cattle and sheep, several were used as food to a small extent.

The young blossoms of *Asclepias fascicularis* (*A. mexicana* of Calif. references) were eaten raw by the Yokia Indians of Mendocino County, but not in any large amounts. The taste was said to be sweet and spicy. In some areas the young leaves and stems of *A. speciosa* were used as greens. The flowers were also eaten raw or boiled, and the buds were boiled for soup or with meat.

The most common use for these plants, recorded among almost all the tribes throughout California, was to obtain a kind of chewing gum. The sticky white juice from the pierced or broken stem was stirred and slightly heated until it became solid. Salmon fat or deer grease was then added to give it a more lasting character. If used without the fat or grease, it quickly breaks up. Among the southern tribes it was used without any other ingredient and the solidifying was done by heating or leaving the juice to stand until curdled. The sap from at least four different Milkweeds was used for chewing gum of this kind (*A. speciosa, A. fascicularis, A. erosa, A. eriocarpa*).

A more important use for several Milkweeds (notably *A. speciosa* and *A. eriocarpa*) was to supply tough fibers for making cords and ropes, and for weaving a coarse cloth. The fibers taken from the stems, which were collected early in winter, were sometimes mixed with the fibers of Indian-Hemp, *Apocynum cannabinum.* The stems do not appear to have been soaked to release the

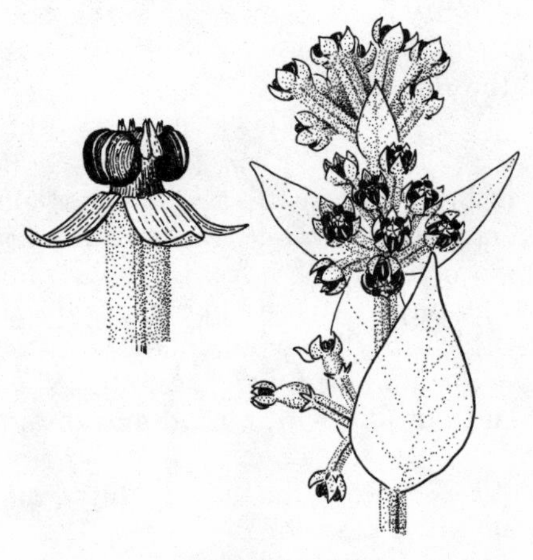

Milkweed

fibers, as with flax, but the bark was removed and the fibers released by first rubbing between the hands and then drawing the mass of fibers over a stock. The cord was rolled on the thigh in the same manner as described under Mescal (p. 19).

The sticky juice of *A. speciosa* was used by some of the desert tribes as a cleansing and healing agent for sores and cuts, also as a cure for warts and ringworm. The silky hairs were burned off the ripe seeds which were then ground and made into a salve for sores. Seed was also boiled in a small amount of water and the liquid used to soak rattlesnake bites to draw out the poison. A hot tea made from the roots was given to bring out the rash in measles or as a cure for coughs. It was also employed as a wash to cure rheumatism. The mashed root, moistened with water, was used as a poultice to reduce swellings.

The juice from the stems of *A. fascicularis* was used as a gum by the Coahuila Indians of the western Colorado Desert and the San Jacinto region. It was also said to have been used to some extent in mounting jewelry. The sticky juice of *A. eriocarpa* was used in Mendocino County as a lotion to make the patterns of tattoo marks, holding the soot while it was being pricked into the skin.

Sea-Blite (*Suaeda californica*)

Sea-Blite is to be found in the salt marshes along the coast from central California to Lower California. The Indians made a rich black dye from the plant for coloring the strands which they used in making the patterns in their coiled baskets. The whole plant was steeped in water, and the dye obtained was said to have been very penetrating and durable with a fetid and disagreeable odor.

Sea-Blite was also one of the many plants from which the Indians gathered seed for pinole (p. 6).

The leaves were gathered, boiled, and eaten as

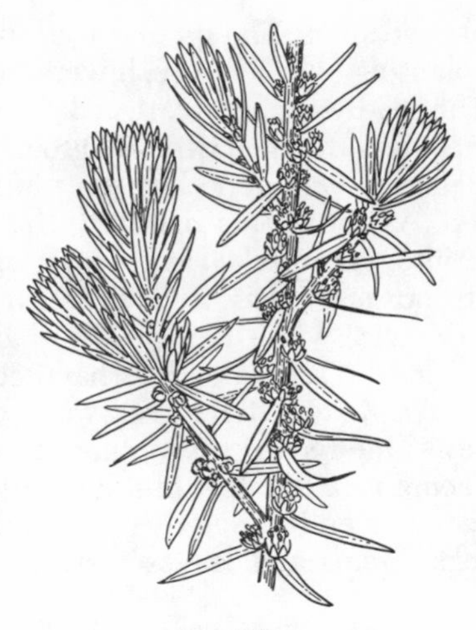

Sea-Blite

greens. In the early days at San Diego the plant was known as Glasswort because of the glassy brittleness of the stem. Gathered in large quantities, the plants were burned and the ashes were used in making soap.

Indian Tobacco (*Nicotiana*)

Indian Tobacco is found growing in the washes, on dryish plains and mesas, and in open valleys throughout a large part of California, below 8000 feet. There are at least three kinds which were used as smoking tobacco by the Indians: *N. Bigelovii, N. attenuata,* and *N. glauca.* It would appear that the practice of smoking was more general in northern than in southern and eastern California. Smoking was really more a "cult," particularly among the tribes of the lower Klamath area. In the Karok economy, smoking was not practiced for pleasure but always for some definite end: as a part of the day's routine, or as a rite prescribed by the tribal customs.

The Karok Indians planted tobacco seeds, *N. Big-*

elovii, in selected spots. The ground was not cultivated, but before planting, logs and brush were burned on the "garden." The seed was then scattered over the cleared area and brush was dragged over the ground to "sweep" it in. No irrigation was done but the plots were carefully weeded. As the plants matured, the leaves were gathered at intervals, packed with care, and wrapped in bracken fronds and twigs of Douglas-Fir so that they would not dry out while being carried down to the village to be dried. Different tribes handled the drying in various ways. Often the leaves were dried in the "sweathouses." One record shows that the tobacco was dried by placing it in the dew in the morning and then taking the leaves in and drying them, repeating this over several weeks. Stems and leaves were harvested sep-

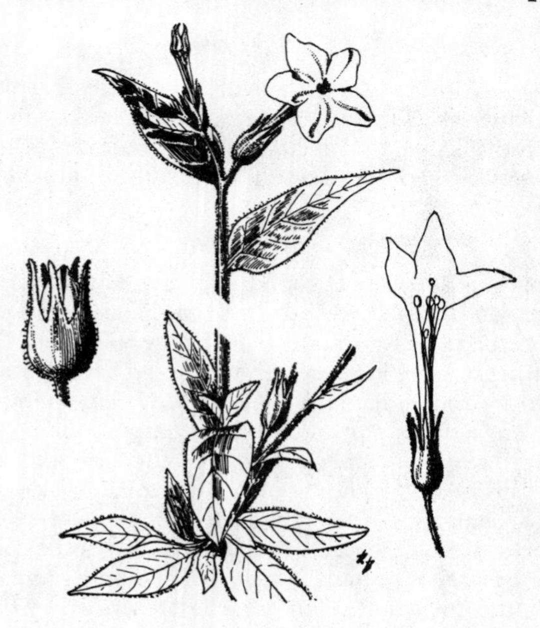

Indian Tobacco

arately, and the different parts of the plant made separate qualities of tobacco, each named and kept apart. The stems made an inferior tobacco which was used for such purposes as gifts to the "Spirits" and in charms and incantations. It was also given to guests of inferior (poor) standing, not as a sign of disrespect, but because it was the custom.

Each year some seeds were gathered from the garden plants, though never from the wild plants which grew about the villages or along the streams. These seeds were cut from the tops of the stems while still green, tied in small bunches, and hung in the house all winter, blackened with the smoke from the fires, and taken down only when the planting time came. Then the capsules were crushed and the seeds scattered directly onto the ground prepared for them.

When the harvested leaves were dry, they were rubbed between the palms of the hands and broken into a not-too-fine powder which was stored in especially woven little baskets which hung in the living house. The supply for current use was carried in the same buckskin bag which held the pipe.

Pipes were made of wood or soapstone, or sometimes of wood with a soapstone bowl. The Karok Indians of the Klamath River used largely Arrowwood, *Philadelphus Lewisii* var. *Gordonianus,* because of the soft pith running through the stem. Usually the pith was bored or punched out with a horn or bone tool, but a curious method was sometimes used. The lengths of wood cut for pipe making were stood on end in oil from the dried salmon until the oil had soaked through the pith from both ends. The bowl was then hollowed out and a little of the pith in the pipe stem and the grub of a little beetle which lives in the dried salmon was placed in the cavity and sealed in with pitch. The wood was then hung up in the living place and left for the grub to eat its way through the pith of the pipe stem. This trick was not always successful; sometimes the grub died without completing the job. Pipes so bored seem to have been particularly valued property.

In general, tobacco leaf was used without anything added, but there are records of mixing it with the dried leaves of Bearberry, *Arctostaphylos uva-ursi,* and of a Manzanita, probably *Arctostaphylos patula.* This was said to be for smoking or as snuff, though the latter practice does not seem to have been general with the Indians and probably only came in after longer contact with the whites. It seems possible that the use of other leaves with smoking tobacco also came from the white settlers.

The stem tobacco was cut into small pieces and ground to a powder in a small stone mortar kept especially for that purpose, and neither pestle nor mortar was ever given any other use.

Tobacco was smoked only by the men, or the women "doctors" who, doing the work of men, must do as the men did. It was rarely chewed, though later reports mention this use of it, as with the taking of snuff, only after longer contact with the whites. Smoking was chiefly done after the evening meal, in the sweathouse, before going to sleep. It was a social ritual, and the pipes were passed around the group. A man never let his pipe out of his sight. Occasionally he would stop for a smoke when on a journey or when meeting someone on the trail.

Apart from smoking, tobacco had a number of uses as medicine. As a pain killer it was used for earache and toothache and occasionally as a poultice. It was considered a poison and had considerable use in the practices of "medicine" by the shamans.

The Indian Tobacco, *N. attenuata,* on the deserts and in the south had something of the same record, though there is nowhere any mention of its having been semicultivated by any of the southern tribes. The Coahuila Indians of the San Jacinto region are said to have powdered it up in special small mortars, mixed it with water, and chewed it.

Medicinally, *N. attenuata* had many uses among the desert tribes. The crushed leaves were made into poultices to soothe rheumatic and other swellings and to place on eczema and similar skin infections. The same material was placed along the gums as a cure for toothache. The chewed leaves could be applied to cuts or bound on rattlesnake bites after the poison had been sucked out.

Smoking was said by the desert Indians to be a cure for colds, especially if the tobacco was mixed with the leaves of the small Desert Sage, *Salvia Dorrii,* or the root of Indian Balsam or Cough Root, *Leptotaenia multifida,* the addition of which was thought to be particularly good for asthma and tuberculosis. The introduced Tree Tobacco, *Nicotiana glauca,* which is common in waste places below 3000 ft., is also said to have been used for smoking by both the Indians and whites. Medicinally the leaves were supposed to be good steamed and used as a poultice to relieve a swollen throat, and steamed into the body for those suffering from rheumatism.

PRESENT-DAY USES OF SOME CALIFORNIA PLANTS

The advanced standard of modern living and the vastly increased population of the world have brought tremendous changes in the methods of supplying the increased wants of mankind. The simple habits of the early peoples, where each family handled its own supply, individually collecting food, clothing, and housing materials from the wild, are no longer possible. This change has also meant that many of the native plant resources in California are no longer of economic importance to us. The rapidly expanding population is inevitably destroying much that once supported the life of the Indians.

But much work has been done, and is still being carried on, in research among the native plants, to find valuable material to help supply our present-day needs. Particularly the desert areas of the Southwest have drawn the concentrated attention of chemists seeking drugs of many types for use not only in medicine but also in manufacturing processes, such as making paper, paints, plastics, or cattle-feed. In this search the old Indian uses of plants have frequently pointed the direction of inquiry, for beneath the superstition and mummery which often accompanied their use of plants for healing purposes, there was nearly always a genuine essence in the plant itself which contributed to the cure of the disease under treatment. While a far smaller number of our native plants are now in actual use than was the case even fifty years ago, there are many which contribute substantially to our needs in ways quite unknown to the general public.

One problem to be met in the modern use of wild plants is the vast quantities of the material necessary to make the use practical and profitable. The only satisfactory way to meet this problem is to cultivate the required plant in great quantity. With many of the desert species this is not a simple matter. The state laws protecting almost all wild species in many states, the stock ranchers' natural objection to large scale harvesting of vegetation on their ranges, and private ownership of much land are conditions to be met. Few of the native plants are presently being cultivated commercially, but some instances are proving successful. A great deal of the work being done on the desert plants of the Southwest combines the plants of California with those of Arizona and New Mexico as one geographical unit.

Perhaps the most important native in the Northwestern United States is the Douglas-Fir. This magnificent forest tree provides a tremendous quantity of lumber, "Oregon Pine," for building and for export. From California to British Columbia the wood is made into pulp for paper making. A resin known as "Oregon Balsam" is also made from this wood.

The Western Hemlock, *Tsuga heterophylla*, is spoken of as the most important tree for the production of wood pulp in the Northwest. The great Redwoods, *Sequoia sempervirens*, are also in large demand for building purposes, and much of the waste from this industry goes into wood pulp and various smaller productions. Though used only as firewood in its native America, the Monterey Pine, *Pinus radiata*, has become a most important tree in Australia and New Zealand, where it is used both as lumber for building and for the pulp industry.

As a source of fiber most of the Spanish Bayonet Yuccas have been used extensively from the earliest times. Among the present-day Indians this fiber is still used in handicrafts for making brooms, brushes, ropes, cords, belts, and mats. During the First World War some eight million pounds of burlap and bagging material were made from Yucca fibers. This fiber is now used in making a heavy kraft paper for flashing and weather-stripping. The fiber of the stiff-leaved or Bayonet Yuccas is considerably stronger than that from the soft leaved types such as *Yucca Whipplei*. When the fibers have been taken out, a good deal of saponin can be obtained from the pulp which is left. The residue from this process can be used as a valuable feed for livestock. There is generally a greater amount of saponin, or soap, in the root of the Yucca, and these are being processed, to a small extent, to produce the lathering material in some modern detergents. During wartime the light, strong wood from the Joshua Tree was used in making splints. It was also used for the banding of apple trees. Fortunately, these uses have been discontinued, for the Joshua Tree is very slow-growing, and any extensive use of the wood might seriously reduce the numbers of this historic species. A note of interest tells of a perfume being made commercially from the flowers of *Yucca baccata*. The skeleton-like wood of some of the Chollas, round-stemmed *Opuntia* species, is being used quite extensively for dried flower arrangements and in the making of curios and souvenirs. This practice must be discouraged because of the severe reduction in the number of these plants which such usage inevitably brings.

Canaigre, *Rumex hymenosapalus*, also known as Wild Rhubarb or Tanner's Dock, has long been a practical source of tannin. The common name comes from the Spanish, *Cana Agria*, "Sour Cane." As far back as the late 1880's and early 1900's roots were collected in the wild and shipped to the East and to Europe for commercial uses. Extensive use of the plant will depend upon the possibility of growing it in sufficient quantity and of handling and extracting the tannin economically. The root contains 25% tannin. Sugar and starch are also present but have not been commercially extracted. A mustard-colored dye is one of the by-products which is being used.

At the present time Tanbark-Oak is much more important in the tannin industry than Canaigre. The second growth of this beautiful forest tree is said to be more desirable than the first growth trees. The leather industry finds properties in the tannin extracted from this wood which makes it especially valuable in that work, particularly in the preparation of sole leather. It is being used to replace Chestnut tannin. At present no extensive commercial enterprise is carried on in California for the use of the wood, though the short-fibered pulp from this second growth wood makes very fine grades of paper.

The large, edible nuts of Jojoba have received almost more attention from investigators and chemists than any other plant of the Southwest. The name Jojoba comes from the original Indian Hohohwi. The oil, which is easily squeezed out in any seed-oil press, is used both in industry and medicine. Though it is said to have no food value, it has been used as a cooking oil. Chemically the oil is a liquid wax which can be readily turned into a hard, white wax. It is colorless and needs very little refining to bring it to maximum purity. The character is not altered by repeated heating to temperatures as high as 285°C, nor does it turn rancid. It has qualities which make it a substitute for sperm oil, hence it is used as a lubricant for delicate machinery. It is also used in the manufacture of shoe and furniture polishes and in the preparation of typewriter ribbons. Laboratory tests have shown it to have qualities which make it valuable in plastics, varnishes, and paints. After the oil has been taken out, the remaining meal makes an excellent cattle feed rich in protein. In its native areas the plant is browsed by deer, cattle, sheep, and goats. Jays, crows, pigeons, and many rodents also help to reduce the wild crop.

A very common shrub in waste places almost throughout the state of California, the Castor-Bean or Castor Oil Plant, *Ricinus communis,* is well known as the source of castor oil. A native of Asia and Africa, it has escaped frequently into the waste spaces of the state. Originally it was introduced as an ornamental plant; small quantities of the seeds are now being gathered for the production of the oil. It should be well known also that the seed from which the oil is extracted is quite poisonous in its raw state, and children have died from eating a single bean. One county in California has made the growing of the Castor-Bean a criminal offense on this account, and it has been eradicated from the county as has Marijuana from the state. The name *Ricinus* is derived from the name of the Mediterranean sheep tick which the seed mimics admirably.

Rush seats for chairs are manufactured in large quantities from the leaves of the Cat-Tail, *Typha latifolia,* and the leaf-sheaths of *Typha glauca* are still used in calking nearly all the watertight, wooden barrels now being made. The seeds contain a drying oil. The down from the fruiting spikes is used as an insulating material for soundproofing and heat insulation. It is also used for buoyant filling in life-jackets. Modern investigations have gone into the use of these plants as substitutes for cotton, wool, and jute, and in the manufacture of paper. The lack of efficient methods of harvesting and processing the material have so far retarded more extensive use of these abundant plants.

Well known to the Indians long before the white man came to the west coast, the bark of the Cascara, *Rhamnus purshiana,* has found a permanent place in the drugs of America. The Cascara is a shrub or small tree of the forests of the Northwestern states, including northern California. In earlier times the bark was gathered from plants in their native forests and weeks or months were often given to searching out the trees. The bark was peeled by the laborious process of handcutting and was carried out to market on the backs of men or horses. In the Northwestern states and Canada cultivation of these trees has been started during the last thirty years, and though the bark is still being sought in the wild, it is likely that commercial needs will be met much more economically in the future.

Far more common in California and extending from north to south is the Coffee Berry, *Rhamnus californica*, a close relative of the true Cascara. The bark of this shrub has similar properties and is frequently used by individuals as a substitute for the genuine article, especially in those areas where the other species is not to be found. The bark of several other native shrubs is being used medicinally to a lesser extent. Deer Brush, *Ceanothus integerrimus*, is used in the preparation of a tonic, Redbud, *Cercis occidentalis*, has astringent bark which provides remedies for diarrhea and dysentery, and the bark of the Flannel Bush, *Fremontia californica*, has been brewed to relieve irritations of the throat.

The thick, sticky leaves of Yerba Santa provide an aromatic syrup, a liquid extract, which is used as an expectorant, and also to disguise the bitter taste in other drugs. A less commercial use for the southern representative of this genus, *Eriodictyon trichocalyx*, is as a cure for the rash caused by Poison-Oak. A good quantity of the fresh leaves is boiled in water to make a very concentrated extract which is painted onto the area affected, as hot as can be borne. The rash disappears in a day or two after a single treatment.

Creosote Bush, *Larrea divaricata*, is the most abundant shrub over many hundreds of square miles of the deserts of the Southwestern United States. The Indians of many tribes used the plant for treating numerous ailments. Today chemists obtain a remarkable drug which is used commercially to delay or prevent butter, oils, and fats from turning rancid. This drug is being produced in large quantities from the leaves and twigs of the plant. When the resins have been taken out, the leaf residue is excellent as feed for livestock, containing as much protein as alfalfa.

It is unlikely that a plant which was so all-important to the native Indians as the Mesquite should be ignored by modern students seeking new sources of material of all kinds. Since the white man came into the Western country the Mesquite has invaded vast areas of grasslands and has become a serious menace to grazing. Research is going on all the time for possible and practical uses to which this plant may be put. The gum, which was used for so many purposes by the Indians, is still being gathered, but in very small quantities. It can be purchased at some drug stores around Tucson, Arizona. Charcoal, which the Indians used a great deal, is still being burned in Mexico and sold in the United States, but no large quantity of this is now available. The chief use for the wood is for fence posts. In a street in San Antonio, Texas, paving blocks of Mesquite wood have withstood the wear and tear of twenty-five years. Investigations point to the possibilities for the use of Mesquite wood in the manufacture of plastics, but that line of search has not yet reached the point of definite products.

The little annual, Chinch-Weed, *Pectis papposa,* which carpets acres of the desert floor with gold after summer rains, matures under good conditions in about six weeks. It is so pungent that when in full growth the desert air is scented with its fragrance, and for that fragrance the flowers were used by the Indians for flavoring foods, especially meats. The women sometimes used it as a perfume, and a golden dye was extracted from the plant. It is now used in the making of an aromatic oil, but attempts at cultivating it under conditions where harvesting would be economical have so far not been very successful.

In supermarkets, on shelves displaying choice spices and seasonings, may be found neat glass jars labeled California Bay-Laurel. Put out by a San Francisco firm, these contain the dried leaves of the California-Bay or Peppernut, *Umbellularia californica.* This leaf is taking the place of the old bayleaf used extensively for flavoring in Europe and the eastern United States. The California-Bay leaves are not easily distinguished from those which they are replacing either in form or taste. They have been used in this manner since the days of the early Spanish settlers, who were not long in discovering this substitute for one of their valued flavorings. Under the name of Oregon Myrtle the wood is carved into bowls, trophies, and souvenirs. The wood is white and fine-grained and works into beautiful artifacts.

There are at least two species of mint, *Mentha spicata,* Spearmint, and *Mentha piperita,* Peppermint, which have escaped into the moist areas along the California coast and are now being dried and packaged. Pennyroyal, *Mentha Pulegium,* may possibly be included with these. All three plants are natives of Europe and came to this country in the transplanted herb gardens of emigrants.

Also from Europe, and cultivated both there and in some areas in the Uuited States, the common Watercress, *Nasturtium officinale,* has strayed off into the streams and is to be found in great quantity in quiet waters and on moist banks in many parts of the state, below about 8000 feet. This healthy salad is harvested and brought into San Francisco and other markets, where it is in constant demand.

Perhaps less generally recognized by sight but well known by name, Horehound, *Marrubium vulgare,* is described by Mr. Munz in his *California Flora* as "a common pestiferous weed in waste places." This weed, brought over from Europe, was highly prized in the days of the old herbalists and is figured as a cure for many ills and ailments by such well-known men as Gerrard and Culpepper. An extract is still used in the preparation of Horehound candy, distributed by a Los Angeles firm. The plant is grown extensively in southern France, and in Norfolk, England, the natives use a drink known as Horehound beer. The chief virtues of the herb are as a cure for coughs and colds, and Horehound candy is an old-time prescription as a cough candy. The name *Marrubium* comes from the Hebrew *marrob,* meaning "a bitter juice."

These notes give only a representative sampling of the California plants which are being used today. In his work, *Native Trees of the San Francisco Bay Region,* in this series of California Natural History Guides, Mr. W. Metcalf lists current uses of many of the woody plants of the state.

REFERENCES

Barrows, David Prescott. The ethno-botany of the Coahuilla Indians of southern California. University of Chicago Press, 1900. 82 pp.

Benson, Lyman, and Robert A. Darrow. A manual of southwestern desert trees and shrubs. University of Arizona Biological Science Bulletin No. 6. Tucson, 1945. 411 pp.

Castetter, Edward F., and Willis H. Bell. Yuman Indian agriculture. University of New Mexico Press. Albuquerque, 1951. 274 pp.

———. The utilization of Mesquite and Screwbean by the aborigines of the American Southwest. University of New Mexico Bulletin No. 314. Albuquerque, 1937. 55 pp.

———. The utilization of Yucca, Sotol and Beargrass by the aborigines of the American Southwest. University of New Mexico Bulletin No. 372. Albuquerque, 1941. 74 pp.

Castetter, Edward F., Willis H. Bell, and Alvin R. Grove. The early utilization and the distribution of Agave in the American Southwest. University of New Mexico Bulletin No. 335. Albuquerque, 1938. 92 pp.

Chesnut, V. K. Plants used by the Indians of Mendocino County, California. Contributions from U.S. National Herbarium 7:295-408. 1902.

Cruise, Robert R. A chemurgic survey of the desert flora in the American Southwest. Economic Botany Vol. 3, (2) pp. 111-131. 1948.

Harrington, J. P. Tobacco among the Karuk Indians of California. Smithsonian Institution, Bureau of American Ethnology, Bulletin 94. 1932. 284 pp.

Havard, Valery. The food plants of North American Indians. Bulletin of Torrey Botanical Club 22:98-123. 1895.

———. The drink plants of the North American Indians. Bulletin of Torrey Botanical Club 23:33-46. 1896.

Henkel, Alice. American drug roots. Bureau of Plant Industry Bulletin 107. Government Printing Office, Washington, D.C., 1907. 80 pp.

Jepson, W. L. A manual of the flowering plants of California. University of California Press. Berkeley, 1925. 1238 pp.

Kroeber, A. L. Handbook of the Indians of California (edition Calif. Book Co. Ltd., 1953). With maps. 995 pp.

Munz, Philip A. and David D. Keck. A California flora. University of California Press. Berkeley and Los Angeles, 1959. 1681 pp.

Palmer, Edward. Plants used by the Indians of the United States. The American Naturalist 12:593-606, 646-655. 1878.

Romero, John Bruno. The botanical lore of the California Indians. Vantage Press Inc. New York, 1954. 82 pp.

Saunders, C. F. Useful wild plants of the United States and Canada. McBride and Co. New York, 1920. 275 pp.

———. Western wild flowers and their stories. Doubleday Doran and Co., Garden City, New York, 1933. 320 pp.

———. With the wild flowers and trees in California. McBride Nast and Co., New York, 1914. 286 pp.

Schenck, Sara M., and E. W. Gifford. Karok ethnobotany. University of California Anthropological Records 13 (6):377-392. 1952.

Stuhr, Ernst T. Manual of Pacific Coast drug plants. 1933. 189 pp.

Train, Percy, J. R. Hendrichs, and W. A. Archer. Medicinal uses of plants by Indian tribes of Nevada. Contributions toward a flora of Nevada No. 33. Bureau of

Plant Industry, Department of Agriculture. Washington, D.C., 1941. 199 pp.
Yanovsky, Elias. Food plants of the American Indians. U.S. Department of Agriculture, Miscellaneous Publications 237. Washington, D.C., 1936. 84 pp.

CHECKLIST-INDEX OF COMMON AND SCIENTIFIC NAMES